tredition®

tredition was established in 2006 by Sandra Latusseck and Soenke Schulz. Based in Hamburg, Germany, tredition offers publishing solutions to authors and publishing houses, combined with worldwide distribution of printed and digital book content. tredition is uniquely positioned to enable authors and publishing houses to create books on their own terms and without conventional manufacturing risks.

For more information please visit: www.tredition.com

TREDITION CLASSICS

This book is part of the TREDITION CLASSICS series. The creators of this series are united by passion for literature and driven by the intention of making all public domain books available in printed format again - worldwide. Most TREDITION CLASSICS titles have been out of print and off the bookstore shelves for decades. At tredition we believe that a great book never goes out of style and that its value is eternal. Several mostly non-profit literature projects provide content to tredition. To support their good work, tredition donates a portion of the proceeds from each sold copy. As a reader of a TREDITION CLASSICS book, you support our mission to save many of the amazing works of world literature from oblivion. See all available books at www.tredition.com.

 Project Gutenberg

The content for this book has been graciously provided by Project Gutenberg. Project Gutenberg is a non-profit organization founded by Michael Hart in 1971 at the University of Illinois. The mission of Project Gutenberg is simple: To encourage the creation and distribution of eBooks. Project Gutenberg is the first and largest collection of public domain eBooks.

Mission Furniture How to Make It, Part I

H. H. (Henry Haven) Windsor

Imprint

This book is part of TREDITION CLASSICS

Author: H. H. (Henry Haven) Windsor
Cover design: Buchgut, Berlin – Germany

Publisher: tredition GmbH, Hamburg - Germany
ISBN: 978-3-8472-3973-4

www.tredition.com
www.tredition.de

Copyright:
The content of this book is sourced from the public domain.

The intention of the TREDITION CLASSICS series is to make world literature in the public domain available in printed format. Literary enthusiasts and organizations, such as Project Gutenberg, worldwide have scanned and digitally edited the original texts. tredition has subsequently formatted and redesigned the content into a modern reading layout. Therefore, we cannot guarantee the exact reproduction of the original format of a particular historic edition. Please also note that no modifications have been made to the spelling, therefore it may differ from the orthography used today.

Mission Furniture

HOW TO MAKE IT

PART I

POPULAR MECHANICS HANDBOOKS

CHICAGO

POPULAR MECHANICS CO.

Copyrighted, 1909,
by H.H. WINDSOR

This book is one of the series of Handbooks on industrial subjects being published by the Popular Mechanics Company.

Like Popular Mechanics Magazine, and like the other books in this series, it is "written so you can understand it."

The purpose of Popular Mechanics Handbooks is to supply a growing demand for high-class, up-to-date and accurate text-books, suitable for home study as well as for class use, on all mechanical subjects.

The text and illustrations, in each instance, have been prepared expressly for this series by well known experts, and revised by the editor of Popular Mechanics.

CONTENTS

- HOME-MADE MISSION CHAIR , 5
- HOW TO MAKE A LAMP STAND AND , 8
- HOW TO MAKE A PORCH CHAIR , 15
- HOW TO MAKE A TABOURET , 17
- HOW TO MAKE A MORRIS CHAIR , 22
- HOME-MADE MISSION BOOK RACK , 27
- HOW TO MAKE A MISSION LIBRARY , 29
- HOME-MADE MISSION CANDLESTICK , 35
- ANOTHER STYLE OF MISSION CHAIR , 36
- HOW TO MAKE AND FINISH A MAGAZINE , 42
- HOME-MADE LAWN SWING , 47
- HOW TO MAKE A PORTABLE TABLE , 50
- HOW TO MAKE A COMBINATION BILLIARD , 51
- EASILY MADE BOOK SHELVES , 56
- A BLACKING CASE TABOURET , 57
- HOW TO MAKE A ROLL TOP DESK , 62
- HOW TO MAKE A ROMAN CHAIR , 67
- HOW TO MAKE A SETTEE , 70
- HOW TO MAKE A PYROGRAPHER'S TABLE , 74
- MISSION STAINS , 76
- FILLING OAK , 77
- WAX FINISHING , 78
- THE FUMING OF OAK , 78
- HOW TO MAKE BLACK WAX , 78
- THE 40 STYLES OF CHAIRS , 80
- HOW TO MAKE A PIANO BENCH , 87
- HOW TO MAKE A MISSION SHAVING , 89
- A MISSION WASTE-PAPER BASKET , 93
- A CELLARETTE PEDESTAL , 96
- A DRESSER , 100
- A MISSION SIDEBOARD , 103
- A HALL OR WINDOW SEAT , 107
- A MISSION PLANT STAND , 109

- **A BEDSIDE MEDICINE STAND**, 112
- **A MISSION HALL CHAIR**, 115

LIST OF ILLUSTRATIONS

- Suitable for Dining Room Use, 5
- Details of Chair Construction, 6
- The Completed Lamp, 9
- Construction of Shade, 11
- Details of Construction of Library Lamp Stand, 12
- Details of Home-Made Porch Seat, 14
- Porch Chair Finished, 16
- Details of Tabouret, 18
- Tabouret as Completed, 20
- Complete Morris Chair Without Cushion, 23
- Details of a Morris Chair, 24
- Light but Strong, 27
- Details of Stand, 28
- This Picture is from a Photograph of the Mission Table Described in This Article, 29
- Showing Dimensions of Table, 30
- Details of Table Construction, 32
- Candlestick, 35
- Details of Candlestick, 35
- Mission Chair Complete, 37
- Details of Mission Chair Construction, 39
- Completed Stand, 43
- Details of the Magazine Stand, 45
- The Completed Swing, 47
- Details of Seat, 48
- Showing Construction of Stand, 49
- Table for Outdoor Use, 50
- By Swinging the Top Back the Table is Transformed into the Elegant Davenport Seen on the Opposite Page, 52
- The Billiard Table as Converted into a Luxurious Davenport — A Child Can Make the Change in a Moment, 53
- Details Showing Dimensions of Parts, 54
- Details of Shoe Rest, 56

- Details of Tabouret Construction, 57
- The Desk Complete, 58
- Details of Tabouret Construction, 59
- The Desk Complete, 61
- Rolltop Details, 62
- Details, 64
- Detail of Pigeonholes, 66
- The Roman Chair, 67
- Details of Parts of Chair, 69
- A Complete Two-Cushion Settee, 71
- Details of a Mission Settee, 72
- Details of the Cushion, 73
- Convenient Pyrographer's Table, 74
- Storage for Apparatus, 75
- Chairs 1, 81
- Chairs 2, 83
- Chairs 3, 85
- Chairs 4, 86
- Piano Bench, 87
- Piano Bench Details, 88
- Shaving Stand Details, 90
- Shaving Stand Complete, 91
- Mirror Frame and Standards Details, 92
- Waste-Paper Basket to Match Library Table, 93
- Detail of Waste-Paper Basket, 94
- Plain-Oak Cellarette Pedestal, 97
- Detail of Cellarette Pedestal, 99
- Dresser in Quarter-Sawed Oak, 101
- Detail of the Dresser, 102
- Detail of the Mission Sideboard, 104
- Mission Sideboard in Quarter-Sawed Oak, 105
- Seat Made of Quarter-Sawed Oak, 107
- Detail of the Hall or Window Seat, 109
- Detail of the Plant Stand, 110
- Complete Plant Stand, 111
- Medicine Stand in Quarter-Sawed Oak, 113

- Detail of the Medicine Stand, 114
- Detail of the Hall Chair, 116
- Complete Hall Chair in Plain Oak, 117

HOME-MADE MISSION CHAIR

Suitable for Dining Room Use

[6]

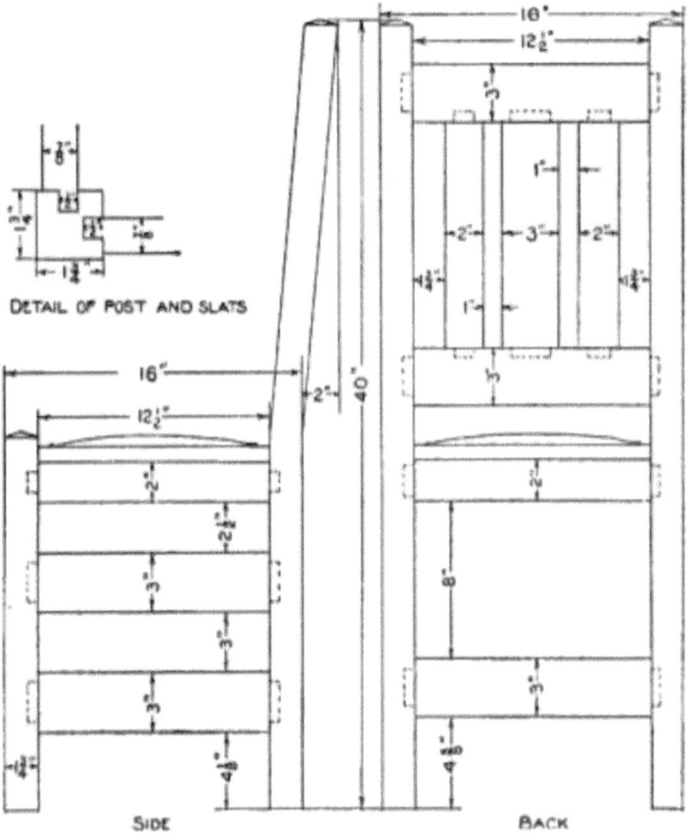

DETAIL OF POST AND SLATS

SIDE

BACK

Details of Chair Construction

A mission chair suitable for the dining room can be made from any one of the furniture woods to match the other articles of furniture. The materials can be secured from the planing mill dressed and sandpapered ready to cut the tenons and mortises. The material list can be made up from the dimensions given in the detail drawing. The front legs or [7] posts, as well as the back ones, are made from 1-3/4-in. square stock, the back ones having a slope of 2 in. from the seat to the top. All the slats are made from 7/8-in. material and of such widths as are shown in the detail. The three upright slats in the back are 3/4-in. material. The detail drawing shows the side and back, the front being the same as the back from the seat down. All joints are mortised in the posts, as shown. The joints, however, can be made with dowels if desired. If making dowel joints they must be clamped very tight when glued and put together. The seat can be made from one piece of 7/8-in. material, fitted with notches around the posts. This is then upholstered with leather without using springs. Leather must be selected as to color to suit the kind of wood used in making the chair. The seat can also be made with an open center for a cane bottom by making a square of four pieces of 7/8-in. material about 4 in. wide. These pieces are fitted neatly to the proper size and dowelled firmly together. After the cane is put in the opening the cane is covered over and upholstered with leather in the same manner as with a solid bottom.

[8]

HOW TO MAKE A LAMP STAND AND SHADE

A library light stand of pleasing design and easy construction is made as follows: Square up a piece of white oak so that it shall have a width and thickness of 1-3/4 in. with a length of 13 in. Square up two pieces of the same kind of material to the same width and thickness, but with a length of 12 in. each. Square up two pieces to a width and length of 3 in. each with a thickness of 1-1/8 in.

If a planing mill is near, time and patience will be saved by ordering one piece 1-3/4 in. square and 40 in. long, two pieces 1-1/8 in. thick and 3 in. square, all planed and sandpapered on all surfaces. The long piece can then be cut at home to the lengths specified above.

The 13-in. piece is for the upright and should have a 1/2-in. hole bored the full length through the center. If the bit is not long enough to reach entirely through, bore from each end, then use a red-hot iron to finish. This hole is for the electric wire or gas pipe if gas is used.

The two pieces for the base are alike except the groove of one is cut from the top and of the other from the under side, as shown. Shape the under sides first. This can best be done by placing the two pieces in a vise, under sides together, and boring two holes with a 1-in. bit. The center of each hole will be 2-1/2 in. from either end and in the crack between the pieces. The pieces can then be taken out, lines gauged on each side of each, and the wood between [10] the holes removed with turning saw and scraper steel.

The Completed Lamp

The width of the grooves must be determined by laying one piece upon the other; a try-square should be used to square the lines across the pieces; however, gauge for depth, gauging both pieces from their top surfaces. Chisel out the grooves and round off the corners as shown in the sketch, using a 3/4-in. radius.

These parts may be put together and fastened to the upright by means of two long screws from the under side, placed to either side of the 1/2-in. hole. This hole must be continued through the pieces forming the base.

The braces are easiest made by taking the two pieces which were planed to 1-1/8 in. thick and 3 in. square and drawing a diagonal on each. Find the middle of this diagonal by drawing the central portion of the other diagonal; at this point place the spur of the bit and bore a 1-in. hole in each block.

Saw the two blocks apart, sawing along a diagonal of each. Plane the surfaces on the saw cut smooth and sandpaper the curve made by the bit. Fasten the braces in place by means of roundhead blued screws.

To make a shade such as is shown in the illustration is rather difficult. The shade is made of wood glued up and has art glass fitted in rabbets cut on the inner edges. Such shades can be purchased ready to attach. The sketch shows one method of attaching. Four small pieces of strap iron are bent to the shape shown and fastened to the four sides of the upright. Electric globes—two, three or four may be attached as shown. [11]

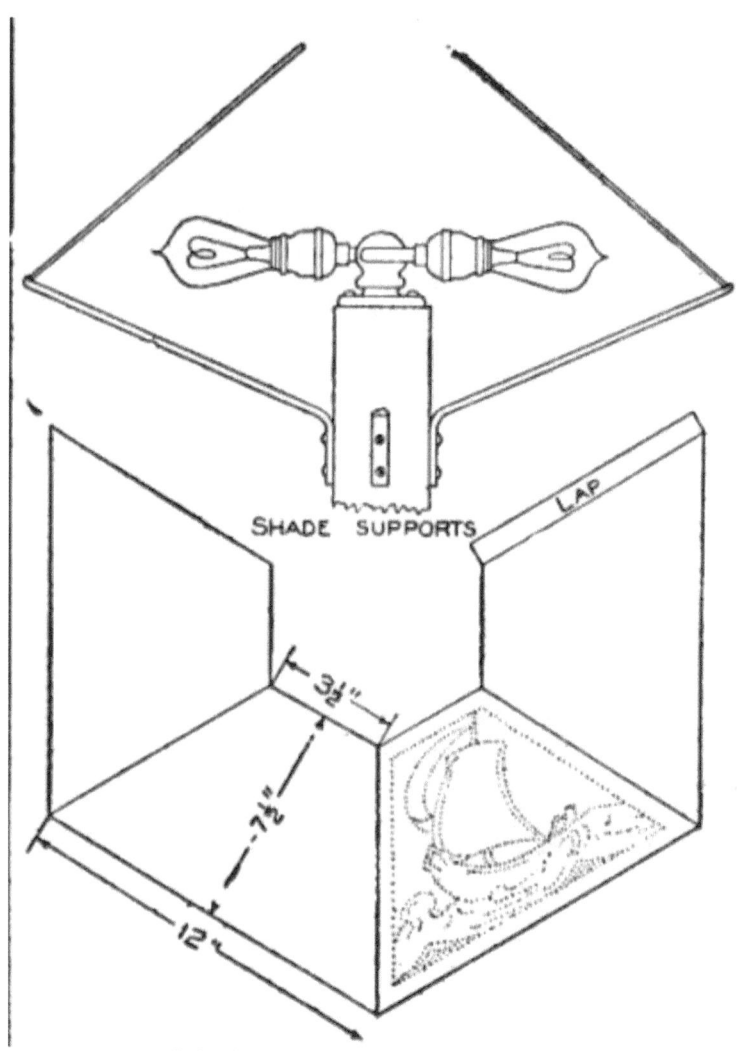

Construction of Shade

The kind of wood finish for the stand will depend upon the finish on the wooden shade, if shade is purchased. Brown Flemish is obtained by first staining the wood with Flemish water stain diluted by the addition of two parts water to one part stain. When this is dry, sandpaper the "whiskers" which were raised by the water and fill with a medium dark filler. Directions will be found on the filler cans. When filler has hardened, apply two coats of wax. [12]

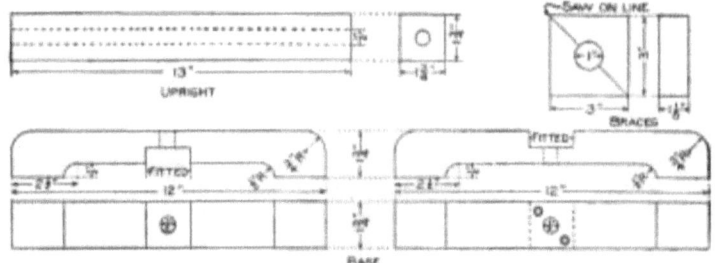

Details of Construction of Library Lamp Stand

[13]

The metal shade as shown in the sketch is a "layout" for a copper or brass shade of a size suitable for this particular lamp. Such shades are frequently made from one piece of sheet metal and designs are pierced in them as suggested in the "layout." This piercing is done by driving the point of a nail through the metal from the under side before the parts are soldered or riveted together. If the parts are to be riveted, enough additional metal must be left on the last panel to allow for a lap. No lap is needed when joints are soldered.

A better way, and one which will permit the use of heavier metal, is to cut each side of the shade separately and fasten them together by riveting a piece of metal over each joint. The shape of this piece can be made so as to accentuate the rivet heads and thus give a pleasing effect.

For art-glass the metal panels are cut out, the glass is inserted from the under side and held in place by small clips soldered to the frame of the shade.

Pleasing effects are obtained by using one kind of metal, as brass, and reinforcing and riveting with another metal, such as copper.

[14]

Details of Home-Made Porch Seat

[15]

HOW TO MAKE A PORCH CHAIR

The illustration shows a very comfortable and attractive porch chair that can be made with few tools and easily procured material. Most any kind of wood will answer, says the American Carpenter and Builder, but if open grained wood, such as oak or chestnut, is used, the parts should be filled with a paste filler. If the natural color of the wood is not desired, the wood may first be stained, the filler being colored somewhat darker than the stain.

Procure enough lumber to make all the pieces shown in the detail drawing and finish to the dimensions shown, being careful to make the corresponding pieces exactly alike in order to preserve the perfect symmetry which is necessary in work of this kind. In boring the holes care must be taken to keep both edges of the holes sharp and clean. The holes should each be bored until the spur shows; the bit should then be withdrawn and the rest of the boring be done from the other side. The semicircular notches are made by placing the two pieces edge to edge in the vise and placing the spur of the bit in the crack. The 1-in. bit is used. As it will be difficult to finish the boring of these blocks from the second side, the parts remaining may be cut out with the knife after the pieces have been separated.

Five 1/2-in. dowel rods are needed. It is possible to get these in one long piece if you happen to live near a mill and then all you will have to do is to saw off the desired lengths. However, if they cannot be got easily you can make your own. Two [16] rods each 18-1/4 in. long; two rods each 20-1/4 in. and one rod 22-1/4 in. give the exact lengths. It is well to cut each piece a little longer than required so that the ends which are imperfectly formed may be cut off. These rods should fit tight and may be fastened in addition with a small screw or nail from the under or back side.

Porch Chair Finished

[17]

The hand rests should be nailed to the arms with small nails or brads before the arms are bolted. The illustration of the assembled chair shows the relative position.

The bolts should be 1/4 in. and of the following lengths: 4 bolts 2-1/4 in. long; 2 bolts 2 in. long; 2 bolts 3 in. long. Washers should be placed between adjacent pieces of wood fastened together with bolts and also at both ends of the bolts. This will require 26 washers in all. While the size of the chair may be varied, it will be necessary to keep the proportions if the parts are to fold properly.

HOW TO MAKE A TABOURET

Secure from the planing mill the following pieces and have them planed and sandpapered on two surfaces: For the top, one piece 7/8 in. thick and 17 in. square. For the legs, four pieces 7/8 in. thick, 4-3/4 in. wide and 18-1/2 in. long. For the lower stretchers, two pieces 7/8 in. thick, 2-3/4 in. wide and 15-3/4 in. long. For the top stretchers, two pieces 7/8 in. thick, 2-1/4 in. wide and 13-1/4 in. long. No stock need be ordered for the keys, as they can be made out of the waste pieces remaining after the legs are shaped.

Begin work on the four legs first. While both sides of each leg slope, it will be necessary to plane a joint edge on each leg from which to lay out the mortises, grooves and to test the ends. It will be necessary to have a bevel square to use in marking off the slopes and for testing them. To get the setting for the bevel square, make a full sized "lay out" or drawing of the necessary lines in their proper relation to one another and adjust the bevel to those lines. [18]

From the joint edge lay out the mortises, grooves and the slopes of sides and ends of the legs. Cut the mortises and grooves first, then shape up the sides. Saw the sides accurately and quite close to the lines, finishing
with
the steel cabinet scraper.

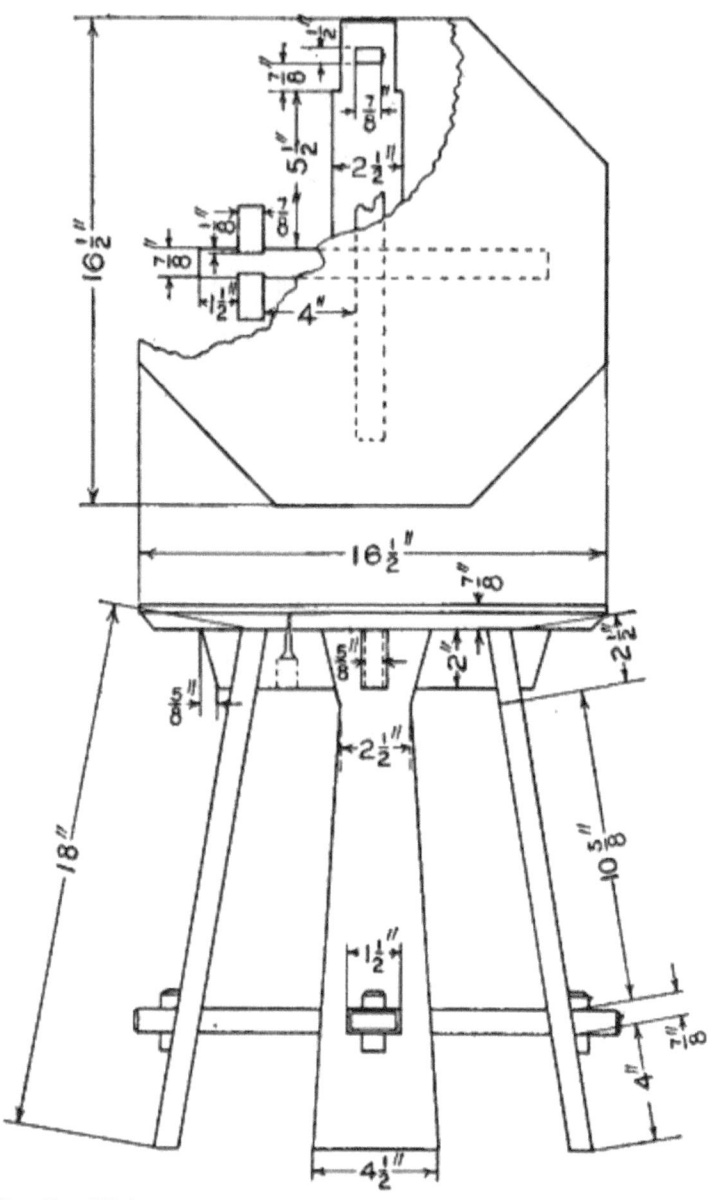

Details of Tabouret

Next make the bottom stretchers. In laying out the cross lap joint, the working faces are both to be up when the joint is completed, therefore lay off one groove on the face of one piece and on the side opposite the face on the other. In gauging for depth, however, be careful to keep the gauge block against the working face of each piece.

In laying out the mortises for the keys, the opening on the top surface is to be made 1/8 in. longer [19] than on the under surface. The slope of the key will therefore be 1/8 in. of slope to each 7/8 in. of length. The drawing shows the mortise as 7/8 in. from the shoulders of the tenon. This distance is the same as the thickness of the leg and to insure the key's pulling the shoulder up against the leg firmly, should any of the legs happen to be a little less than 7/8 in., it is well to make the mortise slightly nearer the shoulder than 7/8 in.

It is a good plan to lay out the mortise in the tenon at the same time the shoulders of the tenons are laid out. Otherwise the joint edge being cut off in making the tenon there is no convenient way to locate this mortise accurately.

Lay off the top stretchers according to the dimensions shown in the drawing. Observe the same precautions about the cross lap joint as were given for the lower stretchers, except that the joint edges are to be placed up in this latter case. Make sure the grooves are laid out in the middle before cutting. As a test, place the pieces side by side, examine the markings, then turn one of them end for end and again examine.

The grooves into which the legs pass are 1/8 in. deep and must be very carefully cut. Their purpose is to give rigidity to the tabouret frame. Bore two holes in each stretcher for the screws that are to fasten the top in place.

Make the keys, scrape all the parts and sandpaper those that were not so treated at the mill. Use glue to fasten the tops of the legs to the top stretchers and assemble these parts.

The top is octagonal or eight-sided. To make it, square up a piece to 16-1/2 by 16-1/2 in. Measure the [20] diagonal, take one-half of it and measure from each corner of the board each way along the edges to locate the places at which to cut off the corners. Connect these points, saw and plane the remaining four sides. There is to be a 5/8-in. bevel on the under side of the top. Scrape and sandpaper these edges and secure the top to the stretchers with screws.

Tabouret as Completed

Much time can be saved and a better result obtained if the wood finishing is done before the parts are put together. Especially is this true if stain and filler are used.

A very pretty finish and one easily put on even [21] after the parts are put together is obtained as follows: Take a barrel and stuff up the cracks or paste paper over them so as to make it as near airtight as possible. In some out-of-the-way place put a dish with about 2 oz. of strong ammonia. Set the tabouret over this dish and quickly invert the barrel over the tabouret. Allow the fumes to act on the wood for at least 15 hours. Remove the barrel and allow the fumes to escape. Polish with several coats of wax such as is used upon floors. Directions for waxing will be found on the cans that contain the wax.. This produces the rich nut-brown finish so popular in Arts and Crafts furniture and is known as fumed oak.

[22]

HOW TO MAKE A MORRIS CHAIR

The stock necessary to make a morris chair of craftsman design as shown in the engraving can be purchased mill-planed and sandpapered on four sides as given in the following list:

- 4 posts 1-3/4 by 3 by 26 in.
- 2 front and back rails 7/8 by 5-1/2 by 24 in.
- 2 side rails 7/8 by 5-1/2 by 28 in.
- 2 arm pieces 7/8 by 5-1/2 by 37 in.
- 7 slats 3/8 by 2 by 24 in.
- 2 cleats 1 by 1 by 22-1/2 in.
- 2 back stiles 1 by 2-1/2 by 24-1/2 in.
- 2 back rails 1 by 2 by 17 in.
- 3 back slats 3/8 by 1-1/2 by 19 in.
- 1 back support 3/4 by 3/4 by 24 in.
- 2 support rests 1 by 1-1/2 by 8-1/2 in.
- 2 dowels 1/2 in. diameter, 6 in. long.

[23]

Complete Morris Chair Without Cushion

First make and put together the sides of the chair. While the glue is setting on these parts make and assemble the back. The front and back rails may next be made and placed and the cleats and bottom slats fastened. With the adjustment of the back the chair is ready for the finish.

The posts are to be tenoned on the upper ends. These tenons are to project 3/16 in. above the arm and should be slightly beveled. The lower ends of the posts, likewise, all other projecting ends, should be beveled to avoid their splintering. All sharp corners, as on the arms, should be sandpapered just enough to take their sharpness off, so as not to injure the hand.

[24]

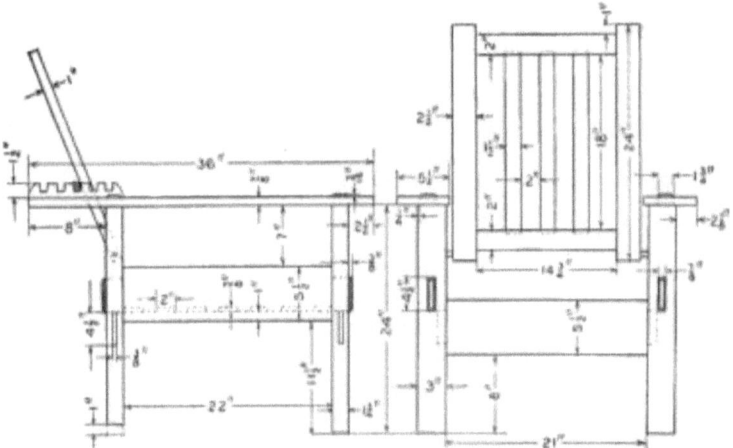

Details of a Morris Chair

That the chair may be properly inclined, the rear posts are cut 1 in. shorter than the forward ones. To get the correct slant on the bottoms of these posts, lay a straightedge so that its edge touches the bottom of the front post at its front surface, but [25] keep it 1 in. above the bottom of the rear post. Mark with pencil along the straightedge across both posts.

At the rear ends of the arms are the notched pieces that allow the back to be adjusted to different angles. These pieces may be fastened in place either by means of roundhead screws from above or flatheads from underneath the arms. The notches are to be cut 3/4 in. deep. If more than three adjustments are wanted, the arms must be made correspondingly longer.

The dimensions for the tenons on all the larger pieces will be found on the drawing. For the back, the tenons of the cross pieces, the rails, should be 3/8 by 1-1/4-in. For the slats, the easiest way is to not tenon them but to "let in" the whole end, making the mortises in the rails 3/8 by 1-1/2 in. This will necessitate cutting the sides of the mortises very accurately, but this extra care will be more than compensated by not having to bother with the cutting of tenons on each end of the three back slats.

To finish the chair, put on a coat of water stain, first removing all surplus glue and thoroughly scraping and sandpapering all the parts that were not so treated at the mill. The color of the stain will depend upon the finish desired, whether golden, mission, etc. Water stains cause the grain of the wood to roughen, so it will be necessary to resandpaper the surfaces after the stain has dried, using fine paper. Next apply a coat of filler colored to match the stain. Directions for its application will be found upon the cans in which the filler comes. After the filler has hardened put on a very thin coat of shellac. [26]

What step is taken next will depend upon what kind of a surface is desired. Several coats of polishing wax may be put on. This is easily done—directions will be found on the cans—and makes the most satisfactory finish for mission and craftsman furniture. It is the easiest to apply. Several coats of shellac or of varnish might be put on instead of wax. Each coat of the shellac should be rubbed when thoroughly dried with curled hair or fine steel wool or fine oiled sandpaper. Rub the first coats of varnish with hair-cloth or curled hair and the last coats with pulverized pumice stone and crude oil or raw linseed oil.

Cushions for the chair can be made at home. They may be made of art leather such as Spanish roan skin and the top and bottom parts fastened together by lacing leather thongs through holes previously punched along the edges of the parts. A very pretty effect is obtained by using thongs of a different but harmonious color. The manner of lacing may be any one of the various laces such as are used in lacing belts or as shoestrings. These cushions may be filled with hair or cotton felt. Denim or burlap may also be used as a covering and are much less expensive than the leather. Lace one side and the two ends, then place filling and finish lacing.

Art leather cushions retail at from $16 to $20 a pair and the denim and burlap at $6 to $9.

The bottom cushion should be made the full size of the chair. The front and back rails extend a little above the slats and thus hold it in place. The back cushion will settle down a little and therefore may be made nearly the full length from the slats to the top of the back.

[27]

HOME-MADE MISSION BOOK RACK

Light but Strong

When making the book rack as shown in the accompanying photograph use quarter-sawed oak if possible, as this wood is the most suitable for finishing in the different mission stains. This piece of furniture is very attractive and simple to construct. [28] The upper shelf can be used for vases or a plant of some kind, while the lower shelves afford ample room for books and magazines.

The slats and legs are fastened to the shelves with 2-in. round-headed brass screws. These can be purchased from any hardware store. One screw is used at each joint of a slat and shelf which calls for 32 screws in all. Holes should be bored into the slats and legs in which to insert the screws. This will keep the wood from splitting. The dimensions are given in the diagram sketch, although these may be changed to suit the requirement of the builder. If no glue is used on the joints when setting up, the rack can easily be taken apart and put in a small bundle for moving.

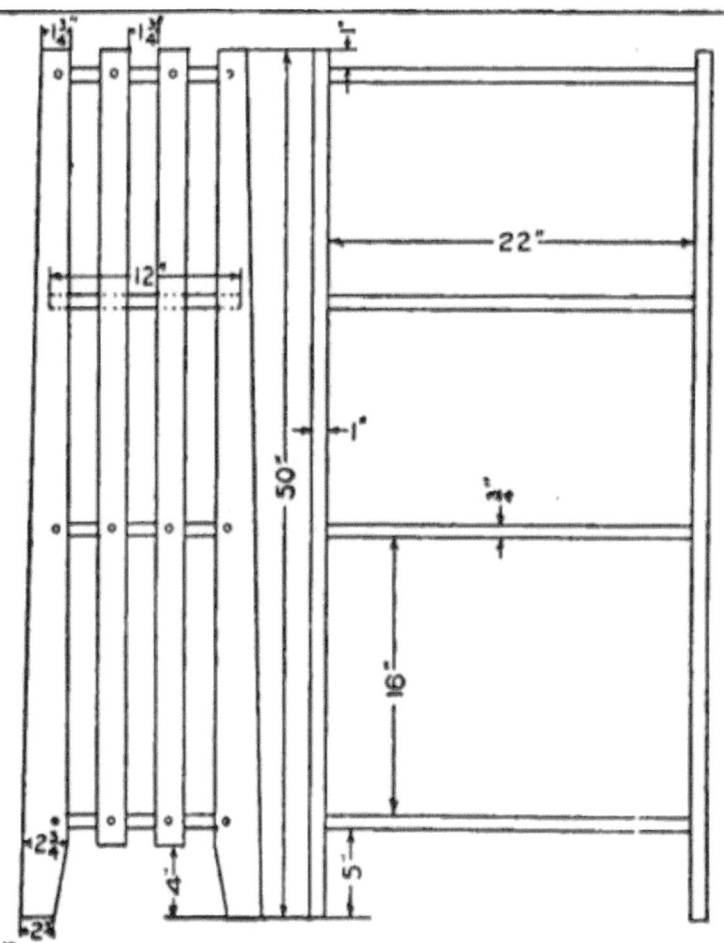

Details of Stand

HOW TO MAKE A MISSION LIBRARY TABLE

The mission library table, the drawings for which are here given, has been found well proportioned and of pleasing appearance. It can be made of any of the several furniture woods in common use, such as selected, quarter-sawed white oak which will be found exceptionally pleasing in the effect produced.

This Picture is from a Photograph of the Mission Table Described in This Article

If a planing mill is at hand the stock can be ordered in such a way as to avoid the hard work of planing and sandpapering. Of course if mill-planed stock cannot be had, the following dimensions must be enlarged slightly to allow for "squaring up the rough."

[30]

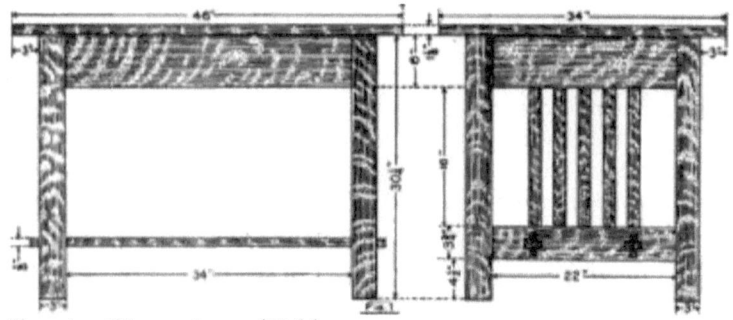

Showing Dimensions of Table

For the top, order 1 piece 1-1/8 in. thick, 34 in. wide and 46 in. long. Have it S-4-S (surface on four sides) and "squared" to length. Also specify that it be sandpapered on the top surface, the edges and ends.

For the shelf, order 1 piece 7/8 in. thick, 22 in. wide and 42 in. long, with the four sides surfaced, squared and sandpapered the same as for the top.

For the side rails, order 2 pieces 7/8 in. thick, 6 in. wide and 37 in. long, S-4-S and sanded on one side. For the end rails, 2 pieces 7/8 in. thick, 6 in. wide and 25 in. long. Other specifications as for the side rails.

For the stretchers, into which the shelf tenons enter, 2 pieces 1-1/8 in. thick, 3-3/4 in. wide and 25 in. long, surfaced and sanded on four sides. For the slats, 10 pieces 5/8 in. thick, 1-1/2 in. wide and 17 in. long, surfaced and sanded on four sides. For the keys, 4 pieces 3/4 in. thick, 1-1/4 in. wide and 2-7/8 in. long, S-4-S. This width is a little wide; it will allow the key to be shaped as desired.

The drawings obviate any necessity for going into detail in the description. Fig. 1 gives an assembly drawing showing the relation of the parts. Fig. 2 gives the detail of an end. The tenons for the side rails are laid off and the mortises placed in the post as are those on the end. Care must be taken, however, not to cut any mortises on the post below, as was done in cutting the stretcher mortises on the ends of the table. A good plan is to set the posts upright in the positions they are to occupy relative to one another and mark with pencil the approximate positions of the mortises. The legs can then [32] be laid flat and the mortises accurately marked out with a fair degree of assurance that they will not be cut where they are not wanted and that the legs shall "pair" properly when effort is made to assemble the parts of the table.

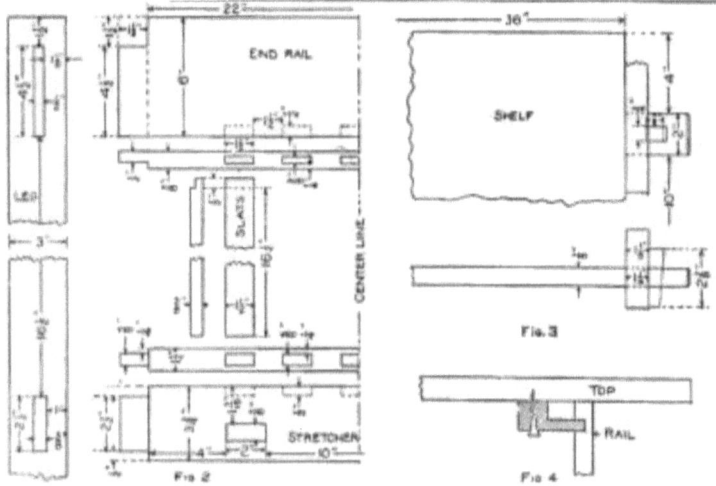

Details of Table Construction

The table ends should be glued up first and the glue allowed to harden, after which the tenons of the shelf may be inserted and the side rails placed.

There is a reason for the shape, size and location of each tenon or mortise. For illustration, the shape of the tenon on the top rails permits the surface of the rail to extend almost flush with the surface of the post at the same time permitting the mortise in the post to be kept away from that surface. Again, the shape of the ends of the slats is such that, though they may vary slightly in length, the fitting of the joints will not be affected. Care must be taken in cutting the mortises to keep their sides clean and sharp and to size.

In making the mortises for the keyed tenons, the length of mortise must be slightly in excess of the width of the tenon—about 1/8 in. of play to each side of each tenon. With a shelf of the width specified for this table, if such allowance is not made so that the tenons may move sideways, the shrinkage would split the shelf.

43

In cutting across the ends of the shelf, between the tenons, leave a hole in the waste so that the turning saw or compass saw can be inserted. Saw within one-sixteenth of the line, after which this margin may be removed with chisel and mallet.

In Fig. 3 is shown two views of the keyed tenon and the key. The mortise for the key is to be placed in the middle of the tenon. It will be noted that [34] this mortise is laid out 1-1/16 in. from the shoulder of the tenon while the stretcher is 1-1/8 in. thick. This is to insure the key's pulling the shelf tightly against the side of the stretcher.

Keys may be made in a variety of shapes. The one shown is simple and structurally good. Whatever shape is used, the important thing to keep in mind is that the size of the key and the slant of its forward surface where it passes through the tenon must be kept the same as the mortise made for it in the tenon.

The top is to be fastened to the rails by means either of wooden buttons, Fig. 4, or small angle irons.

There are a bewildering number of mission finishes upon the market. A very satisfactory one is obtained by applying a coat of brown Flemish water stain, diluted by the addition of water in the proportion of two parts water to 1 part stain. When this has dried, sand with No. 00 paper, being careful not to "cut through." Next, apply a coat of dark brown filler; the directions for doing this will be found upon the can in which the filler is bought. One coat usually suffices. However, if an especially smooth surface is desired, a second coat may be applied in a similar manner.

After the filler has hardened, a very thin coat of shellac is to be put on. When this has dried, it should be sanded lightly and then one or two coats of wax should be properly applied and polished. Directions for waxing are upon the cans in which the wax is bought. A beautiful dull gloss so much sought by finishers of modern furniture will be the result of carefully following these directions.

[35]

HOME-MADE MISSION CANDLESTICK

There are many kinds of mission candlesticks, but few of them carry out the mission design throughout. Herewith is illustrated a candlestick which may be made from the various woods that will have the style and lines of mission craft work. The base is made from 1-in. material, 4-1/2 in. square. Two holes are bored and countersunk for screws to hold the post and handle. The post is 2-1/4 in. high, bored in one end to fit the size of a candle. The post is covered with a 3/8-in. thick cap, 2 in. square. This, also, is bored to fit the candle. The handle is 3/8 in. thick and 3 in. long with a 3/8-in. square mortise and is notched to fit the base. The wood may be selected to match any other piece of furniture and finished in any of the mission stains.

Candlestick

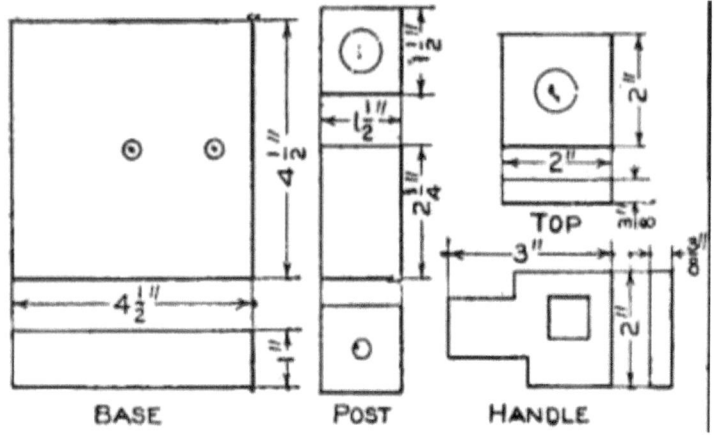

Details of Candlestick

[36]

ANOTHER STYLE OF MISSION CHAIR

The material necessary to make a mission chair as shown in the accompanying illustration may be secured from a planing mill with all four surfaces squared and sandpapered. The mill can do this work quickly and the expense will be nothing compared with the time it takes to do the work by hand.

The following is the stock list:

- 4 legs, 2-1/2 by 2-1/2 by 32-1/2 in.
- 2 bottom end rails 7/8 by 5-3/4 by 23-1/2 in.
- 2 bottom side rails 7/8 by 5-3/4 by 28-1/2 in.
- 2 top end rails 7/8 by 4-1/2 by 23-1/2 in.
- 1 top back rail 7/8 by 4-1/2 by 28-1/2 in.
- 2 cleats 7/8 by 2-1/2 by 26-1/2 in.
- 7 slats 1/2 by 2 by 24 in.

This design was purposely made simple. If it is considered too severe and the worker has had some experience in woodwork, it can easily be modified by adding vertical slats in back and sides. These should be made of 1/2-in. stock and their ends should be "let into" the rails by means of mortises.

Either plain red oak or quarter-sawed white oak will do. Begin by squaring up one end of each leg, marking and cutting them to length and planing up the second ends so that they shall be square. Both the top and bottom of each leg should be beveled or rounded off about 1/4 in. so that they may not splinter or cause injury to the hand.

When all of the legs have been made of the same length, set them on end in the positions they are to have relative to one another and mark with pencil the approximate locations of the mortises. Next, place them on the bench, side by side, even the ends [37] and square sharp lines across to indicate the ends of the mortises. The drawing shows the dimensions to use. A sharp pencil should be used for this marking and the lines should be carried entirely across the two faces of each piece.

Mission Chair Complete

Set the gauge for the side of the mortise nearest the face edge. With this setting, mark all the mortises, then set for the second side of the mortise and complete the gauging.

There are two ways of cutting small mortises in common use. One is by using a chisel of a width just equal to that of the mortise. The other is by using a smaller chisel after the mortise has first been bored with the brace and bit. In the first method the cutting is begun at the middle of the mortise where a V-shaped opening is made the full depth of the mortise that is to be. Continuing from the middle, vertical cuts are taken first toward one end and then toward the other. The chips are pried out as the cutting proceeds. In making the last cut this prying must be omitted, otherwise the edge of the mortise would be ruined. It will be necessary to stand so as to look along the opening in order to get the sides plumb.

This method of cutting, when once the "knack" has been attained, will be found much easier, quicker and more accurate for small openings, such as these, than the usual method. The second method, which is the usual one, needs no description. [39]

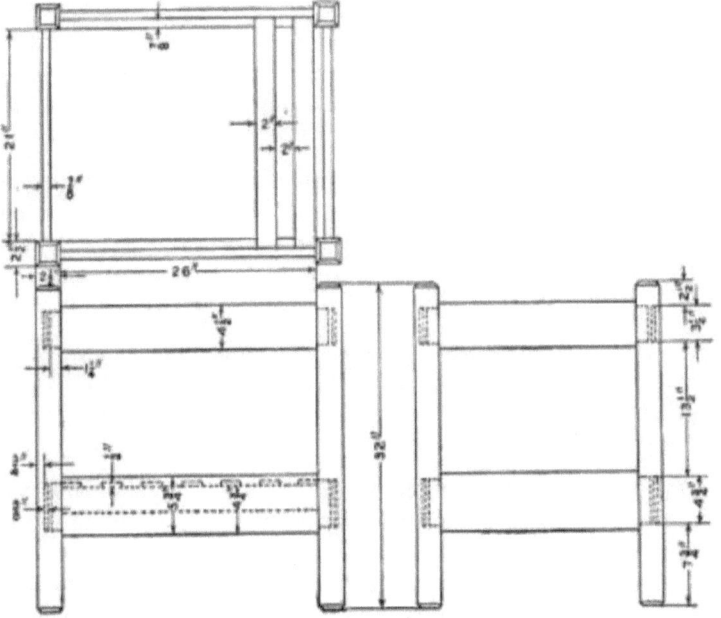

Details of Mission Chair Construction

The rails should next have the tenons cut on their ends. It may not be out of place to remind the amateur that the lengths of the various like pieces can best be laid off by placing them on the bench, measuring off the proper distances on one of them and then with try-square marking across the edges of all of them at once. This not only saves time in that but one set of measurements need be made, but it insures all the pieces being similarly laid off. In measuring off for the shoulders of the tenons, begin at the middle of the length of the rail and measure half of the distance each way. By doing so, if there are any slight differences in the lengths of the pieces this difference will be divided between the two tenons and no harm will be done.

In gauging the tenons take the precaution to mark a working face and joint edge, even if all the surfaces were finish-planed at the mill. It is very important that all tenon gauging be done from these faces. The same is true of the legs or posts, and the slats if there are to be any.

To avoid confusion it is well to number each tenon by means of the chisel with a Roman numeral and its corresponding mortise with the same. This will prevent the fitting of one tenon into more than one mortise.

Put the parts together with warm glue if it can be had, otherwise use the prepared cold glue. In cold weather the wood ought to be warmed before the glue is applied. Put the ends of the chair together first. When the glue has set on these put the other rails in place.

When clamping up the second set of rails make sure the frame of the chair is square. The best way to test for squareness is to measure the diagonals with a stick. Spring the frame until they measure alike, using a brace to hold the frame in position until the glue can harden.

Before staining, scrape off any surplus glue, for [41] stain will not adhere to glue and a white spot will be the result of failing to remove it. Fasten cleats to the front and back rails with screws. To these cleats fasten the slats as shown in the drawing. A cushion of Spanish leather, such as is shown in the photograph, can be bought at the furniture store or the upholsterer's. It can be made by the amateur quite easily, however. The two parts are fastened together with leather thongs and the filling is of hair or elastic felt. A cushion for the back might well be provided.

To finish the wood to match a brown leather proceed as follows: With a cloth or brush, stain the wood with brown Flemish water stain diluted by the addition of four parts of water. When this has dried, sandpaper smooth, using No. 00 paper held on the tips of the fingers. Apply a dark brown filler. When this has flatted, i. e., when the gloss has disappeared, which will be in the course of ten or fifteen minutes, wipe off clean with excelsior and then with waste or a cloth. Allow this to dry over night, then apply two or three coats of wax. Polish each coat with a flannel cloth by briskly rubbing it.

A settle can be made after this design by using longer front and back rails. Rails 42 in. between shoulders will make a good length for a settle.

HOW TO MAKE AND FINISH A MAGAZINE STAND

For the magazine stand shown herewith there will be needed the following pieces:

- 1 top, 7/8 in. by 15-1/2 in. by 16-1/2 in.
- 1 shelf, 7/8 in. by 11-1/2 in. by 12-1/2 in.
- 1 shelf, 7/8 in. by 12-1/2 in. by 14-3/4 in.
- 1 shelf, 7/8 in. by 13-1/2 in. by 16-1/2 in.
- 2 sides, 7/8 in. by 14-1/2 in. by 33-1/2 in.
- 1 brace, 7/8 in. by 3-1/4 in. by 17 in.
- 1 brace, 7/8 in. by 2-1/2 in. by 11-1/2 in.
- 6 braces, 7/8 in. by 2 in. by 2 in.

Order these pieces mill-planed on two surfaces to the thickness specified above and also sandpapered. Quarter-sawed white oak makes the best appearance of all the woods that are comparatively easy to obtain. Plain sawed red or white oak will look well but are more liable to warp than the quarter-sawed. This is quite an element in pieces as wide as these.

Begin work on the sides first. Plane a joint edge on each and from this work the two ends. The ends will be square to the joint edge but beveled to the working face. A bevel square will be needed for testing these beveled ends.

To set the bevel make a drawing, full size or nearly so, of the front view and place the bevel on the drawing, adjusting its sides to the angle wanted. Work from a center line in laying off the drawing.

Having planed the ends, lay off the sides. This is done by measuring from the joint edge along the bottom 14 in., from the joint edge along the top 1-1/2 in. and from this 11 in. Connect the points by means of a pencil and straightedge. [43]

Completed Stand

[44]

Before cutting off the joint edges of the pieces measure off and square lines across to indicate the locations of the shelves. Put both pieces together and mark across both joint edges at once to insure getting both laid off alike.

The design at the bottom can be varied to suit the fancy of the worker. For such a design as is shown, draw on paper, full size, half of it; fold on the center line and with scissors cut both sides of the outline by cutting along the line just drawn. Trace around this pattern on the wood, and saw out with compass or turning saw.

The shelves may now be made. The bevel of the ends of the shelves will be the same as for the ends of the side pieces. The lengths may be obtained by measuring the drawing. Remember that length is always measured along the grain and that the end grain of the shelves must extend from side to side in this stand. The widths may be obtained by measuring the width of the sides at the points marked out on them for the location of the shelf ends. It is best not to have the shelves the full width of the sides, since the edges of the shelves are to be faced with leather. Make each shelf 1/2 in. less than the width of the side, at the place that the shelf is to be fastened.

The top will be squared up in the usual manner, 15 in. wide by 16 in. long.

These parts may now be put together. They may be fastened in any one of a variety of ways. Round-head blued screws may be placed at regular intervals through the sides. Finishing nails may be used and the heads set and covered with putty stained to match the wood. Finish nails may be [45] placed at regular intervals and fancy headed nails used to cover the heads.

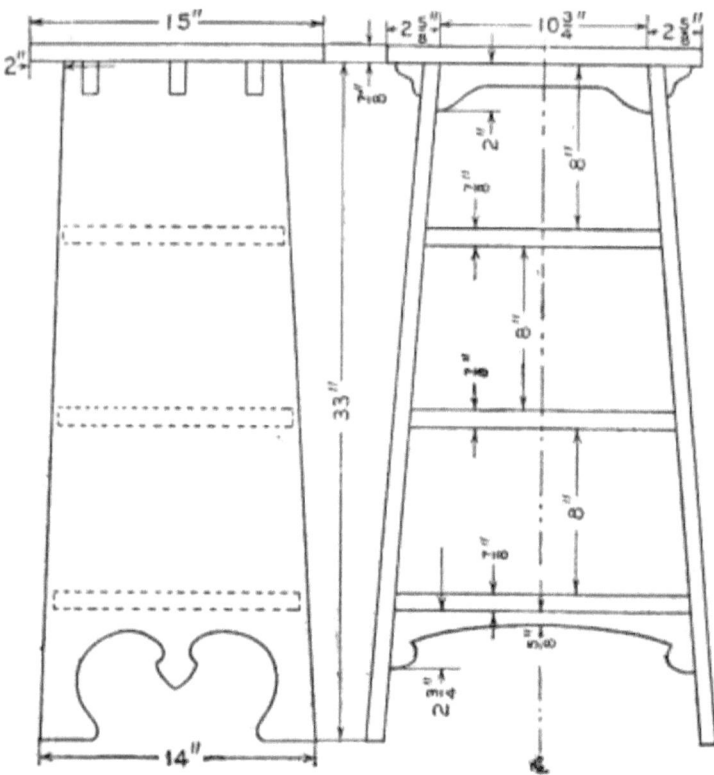

Details of the Magazine Stand

The braces should be formed and fitted but not fastened until the finish has been applied. Thoroughly scrape and sandpaper all parts not already so treated. Probably no other finish appeals to so many people as golden oak. There is no fixed standard of color for golden oak. Different manufacturers have set standards in their part of the country, but [46] the prevailing idea of golden oak is usually that of a rich reddish brown.

Proceed as follows: Egg shell gloss: 1.—One coat of golden oak water stain, diluted with water if a light golden is desired. 2.—Allow time to dry, then sandpaper lightly with fine sandpaper. This is to smooth the grain and to bring up the high lights by removing the stain from the wood. Use No. 00 sandpaper and hold it on the finger tips. 3.—Apply a second coat of the stain diluted about one-half with water. This will throw the grain into still higher relief and thus produce a still greater contrast. Apply this coat of stain very sparingly, using a rag. Should this stain raise the grain, again rub lightly with fine worn sandpaper, just enough to smooth. 4.—When this has dried, put on a light coat of thin shellac. Shellac precedes filling that it may prevent the high lights—the solid parts of the wood—from being discolored by the stain in the filler, and thus causing a muddy effect. The shellac being thin does not interfere with the filler's entering the pores of the open grain. 5.—Sand lightly with fine sandpaper. 6.—Fill with paste filler colored to match the stain. 7.—Cover this with a coat of orange shellac. This coat of shellac might be omitted, but another coat of varnish must be added. 8.—Sandpaper lightly. 9.—Apply two or three coats of varnish. 10.—Rub the first coats with hair cloth or curled hair and then with pulverized pumice stone, crude oil or linseed oil. Affix the braces just after filling, using brads and puttying the holes with putty colored to match the filler. The shelves may be faced with thin leather harmonizing with the oak, ornamental headed tacks being used to fasten it in place.

[47]

HOME-MADE LAWN SWING

The Completed Swing

The coming of spring and summer calls forth various kinds of porch and lawn furniture. A porch or lawn swing to accommodate two or more persons is a thing desired by most people. The lawn swing as shown in the picture is portable and does not need stakes to hold it to the ground. While this swing is substantial and rigid it can be moved from place to place on the lawn, or the chains can be fastened with heavy hooks to the ceiling of a porch instead of using the stand. Either ropes or chains may be used to hang the swing and should be of such length that the seat will be about 20 in. from the ground or floor. [48]

The drawing giving the dimensions for constructing the seat shows how the parts are put together. The front and back apron pieces are mortised to receive a 1-in. square tenon cut on the cross-pieces that support the slats. Each end of the apron pieces extends 4 in., and a hole is bored at A into which the hanging ropes or chains are fastened. If ropes are used, bore the holes to fit the rope and when the end of each rope is put through a hole it is tied in a knot to keep from slipping out. Chains can be fastened with eye bolts. Small carriage or stove bolts are used to hold the slats on the framework and cross pieces. The arm rests are fastened with wood screws.

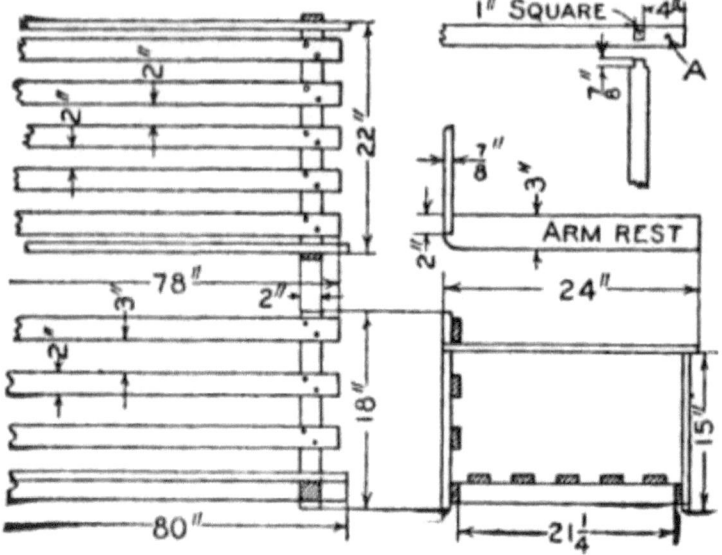

Details of Seat

The drawing for the stand gives all the dimensions for its construction. Split the upright pieces or legs with a saw cut to the length as shown. A bolt should be put through each piece edgewise at the end of the saw cut, to keep the wood from splitting any farther when the ends are spread to receive the bolts through the cross pieces at the top. The upper ends of the ropes or chains are fastened close to and under the bolt holding the inside forks of the uprights. This bolt can be long enough [49] to fasten a clevis that will hang underneath for this purpose. The whole swing can be painted with a forest green color which is very suitable for summer outdoor furniture.

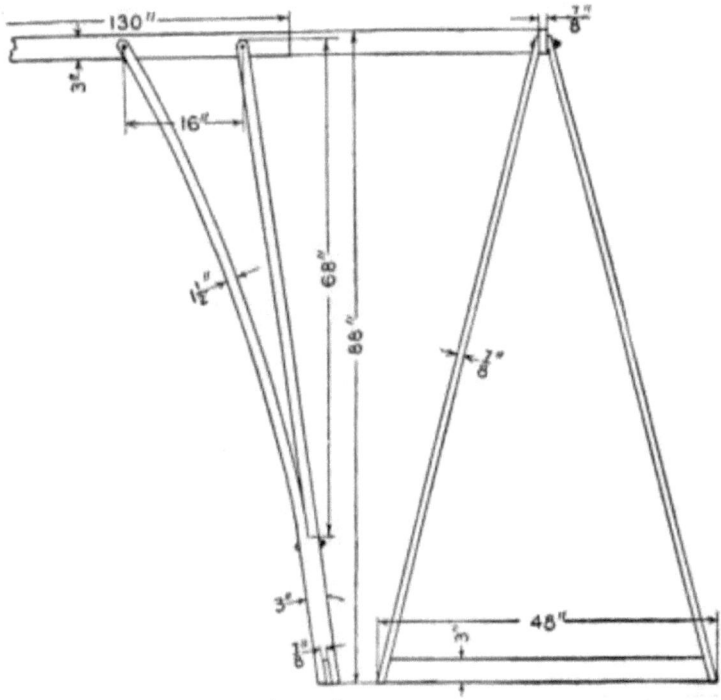

Showing Construction of Stand

[50]

HOW TO MAKE A PORTABLE TABLE

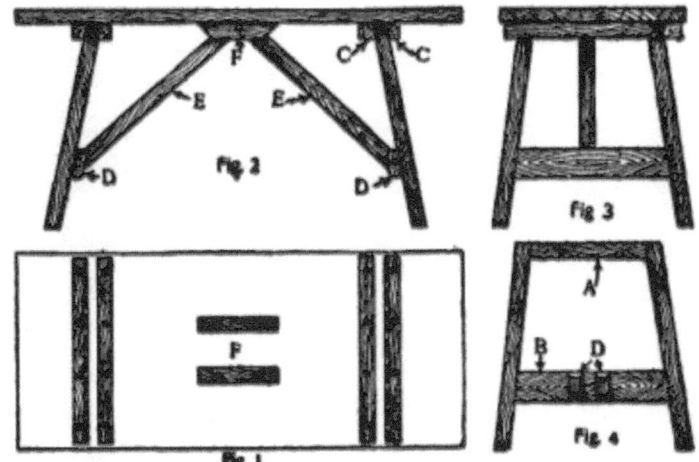

Table for Outdoor Use

A table for outdoor use that can be taken apart, stored or changed from place to place may be made at small expense. Fasten cleats with screws, as shown in Fig. 1, to the bottom of a board of suitable size. The legs are built with a cross piece, A, Fig. 4, at the top which fits into slot formed by the cleats, CC, and a crosspiece, B, that has two cleats, D, making a place to receive the bottom end of the brace, E, Fig. 2. The upper ends of the braces, EE, fit in between two pieces, F, fastened in the middle of the board. The three pins fitting loosely in DD and F, Fig. 2, are all that holds table together. The end view is shown in Fig. 3.

[51]

HOW TO MAKE A COMBINATION BILLIARD TABLE AND DAVENPORT

A small size billiard table which can be converted quickly into a davenport is made as follows: Secure clear, selected plain sawed white oak in sizes as indicated by the drawing. Have these planed at the mill to the widths and thicknesses specified.

The lower part should be made first. Cut the four posts to length, chamfering the ends somewhat so that they will not splinter when in use. Lay out and cut the mortises which are to receive the rails. The lower rails are to be 1-1/8 in. thick and the mortises are to be laid out in the legs so as to bring their outer surfaces almost flush with those of the posts. The upper rails are 2-1/4 in. wide. The slats are 3/4 in. thick. Tenons should be thoroughly pinned to the sides of the mortises as shown in the illustration. The braces are 1-3/4 in. thick and are fastened to place with roundhead screws and glue.

The seat may be made by putting in a solid bottom that shall rest upon cleats fastened to the inner surfaces of the rails. The top of this bottom should rest about 3/4 in. below the top edge of the rails. A well filled leather cushion completes this part.

A more satisfactory result is obtained by putting in springs and upholstering the seat. Upon this the leather cushion can be placed.

[52]

By Swinging the Top Back the Table is Transformed into the Elegant Davenport Seen on the Opposite Page

The Billiard Table as Converted into a Luxurious Davenport—A Child Can Make the Change in a Moment

The top or table is built upon and about a heavy frame of well seasoned 1-3/4-in. by 5-3/4-in. white pine. The parts to this frame are thoroughly mortised and tenoned together. Middle stretchers, lengthwise and [53] crosswise, give added strength and rigidity.

Upon this frame the slate bed is leveled by planing the frame wherever necessary. Slats are fastened to the bed by screws, the heads of which are countersunk so that they may be covered over even with plaster of paris.

[54]

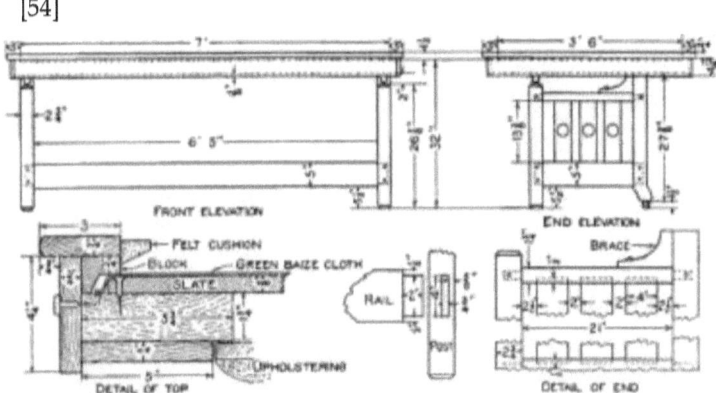

Details Showing Dimensions of Parts

The top and side facings are built together, the angle being reinforced with block and glue, as shown in detail. These facings, to which the cushions are attached, are afterward made fast to the frame by ornamental headed screws. The detail and photograph show the manner of applying the under facing.

Before attaching the top and side facings, the bed cloth should be placed over the slate and fastened. The nap of the cloth should run from the head toward the opposite end of the table. Draw the cloth as tight as possible, taking care that there shall be no wrinkles.

The billiard cushions can be bought ready to cover. The bumpers which keep the top from striking the front posts can be obtained by making proper selection from oak door bumpers carried in stock by hardware dealers. The brass swing bars, most likely, can be obtained at the same place.

The upholstering on the under side of the top—the back of the davenport—is to be built upon a stout frame made of some suitable common wood, and the whole set in the recess formed as shown in the detail drawing—the whole being fastened from the back before the slate bed is put in position.

Effort should be made to select leather of a color that will harmonize with the wood finish which is to be applied.

[56]

EASILY MADE BOOK SHELVES

Very cheap but useful and attractive book shelves are shown in the accompanying drawing. The vertical strips, A, may be 3/4 in. by 2 in. and are screwed to four shelves, B, each cut to the shape of a quarter circle. The screws are all countersunk and as the heads all come on the side next to the wall, they do not show. The design might be varied somewhat to suit the fancy of the builder, although the appearance of the shelves constructed as shown is very pleasing, especially so if the workmanship is good and the wood carefully stained and varnished. The total cost of construction was less than 75 cents.

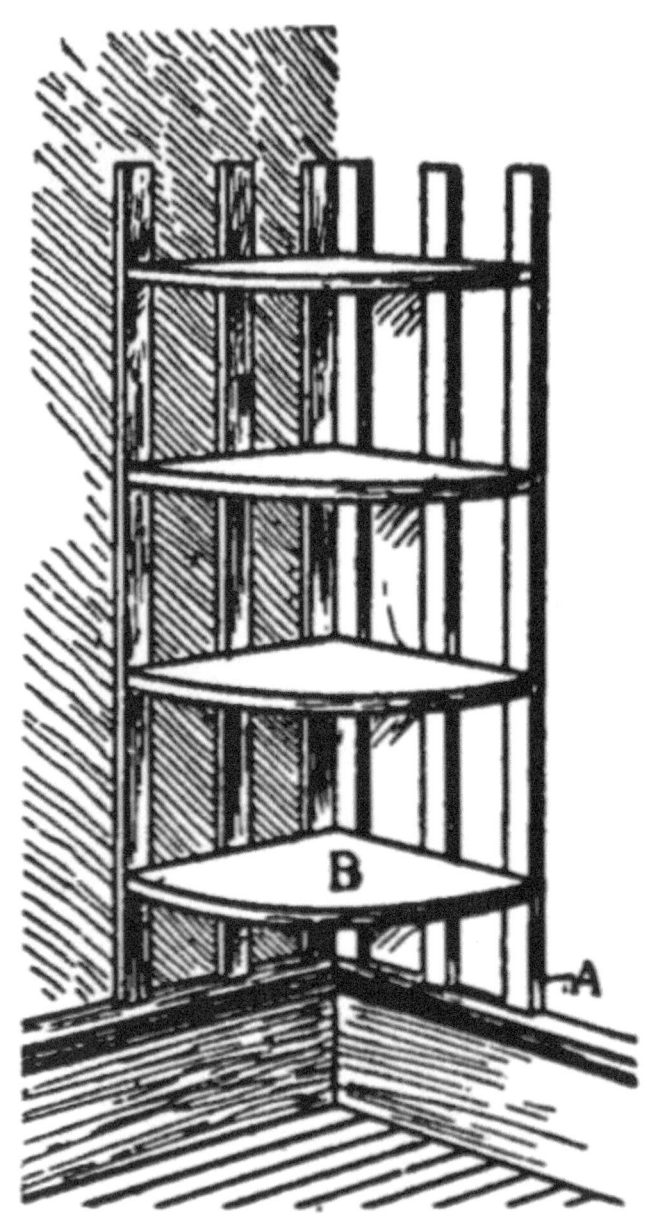

[57]

A BLACKING CASE TABOURET

A substantial piece of mission furniture which may be used as a tabouret or plant stand as well as a blacking case, in which there is a receptacle for brushes, blacking and a shoe rest, is shown in the illustration. The stock can be secured mill-planed, sandpapered and in lengths almost ready to be assembled. The stock list consists of the following pieces:

- 4 posts, 1-1/2 by 1-1/2 by 17 in.
- 4 side rails, 1 by 6-1/2 by 9-1/2 in.
- 2 top pieces, 1 by 8-1/4 by 16-1/2 in.
- 1 bottom, 1/4 by 9-1/2 by 9-1/2 in.
- 1 cleat, 1 by 1 by 18 in.

The posts and cleat are surfaced on four sides, while the other pieces are surfaced on only two [58] sides. The allowance of 1/2 in. on the side rails, top and bottom, is for fitting the joints. Be sure the surfaces of the pieces for the posts are square and the ends sawed square off, making the posts exactly the same length when they come from the mill.

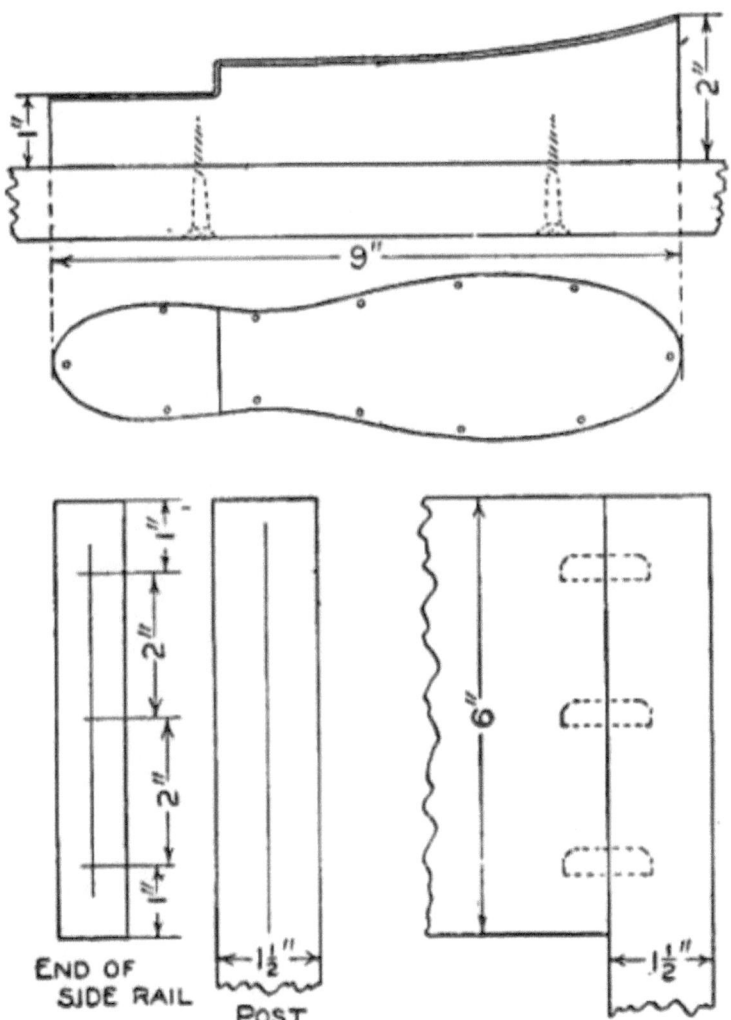

Details of Shoe Rest

Square up the four side rails to 6 by 9 in. Cut one end of each post tapering with a chisel; face and sandpaper the posts and side rails before making the joints. The side rails are attached to the posts with three dowels to each joint. The place for each dowel is located by making a line exactly in the middle lengthwise on each end of each side rail. Three lines are made to intersect this middle line, as shown in the detail. Drive a 1/2-in. brad in each intersection allowing a small portion of each brad to project, and cut off [59] the heads. Gauge a line in the middle of each post at the top where the joints are to be made and press the end of a side rail containing the brads against the post. This will mark the places to bore holes for the dowels. Pull out the brads and bore holes for the dowel pins.

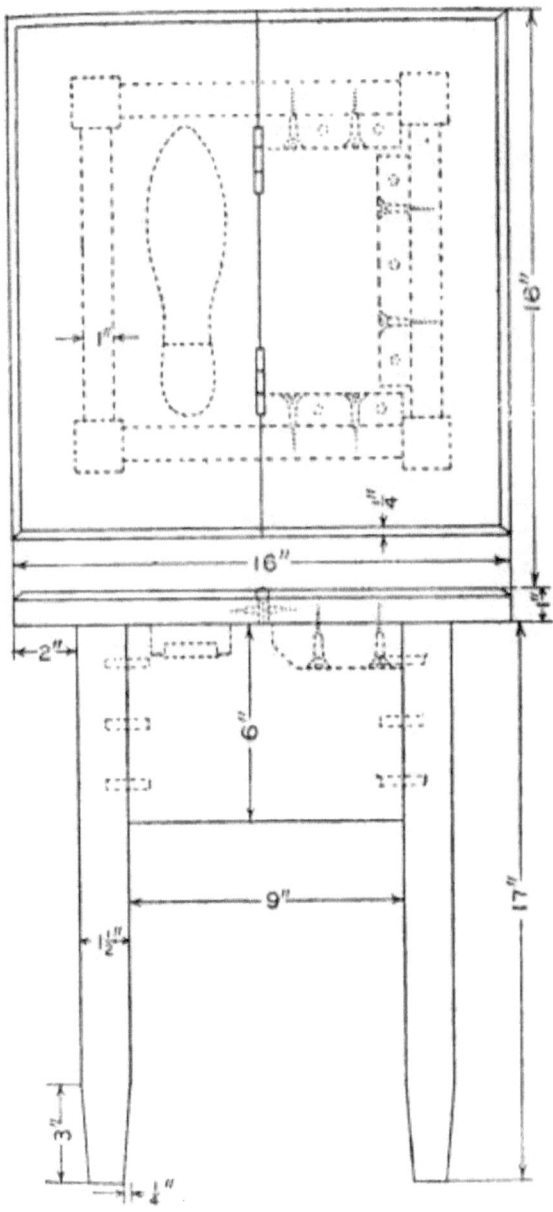

Details of Tabouret Construction

When gluing up the side rails and posts, first put on a coat of glue on the ends of the side rails and let it dry. This will fill up the pores in the end grain of the wood which will make a strong [60] joint when finally glued together. The dowel pins are made 3/8 in. square with a slight taper at the ends. These can be easily forced into the holes, when the ends of the side rails are coated with glue and ready to be put together, by clamps pressing on the outside of the posts.

The bottom is held in position with narrow strips tacked on the lower edge of the side rails. Square up the top pieces to 8 by 16 in. and fasten one piece to the top with cleats and screws as shown in the drawing. The other piece is hinged to the first one with two 2-in. hinges.

The shoe rest can be made from a block of wood and covered with sheet tin, copper or brass, or a cast-iron rest can be purchased. The rest is fastened to the under side of the hinged top. Stain the wood any dark color and apply a very thin coat of shellac. Put on wax and you will have a finish that can be renewed at any time by wiping with a little turpentine and rewaxing.

[61]

HOW TO MAKE A ROLL TOP DESK

The Desk Complete

The materials for this roll top desk can be purchased from a mill dressed and sandpapered so the hardest part of the work will be finished. The wood must be selected to suit the builder and to match other articles of furniture. The following list of materials will be required:

- 68 lineal ft. of 1 by 3 in. hardwood.
- 65 lineal ft. of 1 by 2 in. hardwood.
- 3 lineal ft. of 1/4 by 24 in. hardwood.
- 45 lineal ft. of 1/4 by 10-1/2 in. hardwood.
- 36 lineal ft. of 1 by 12 in. hardwood.
- 35 lineal ft. of 3/8 by 9 in. soft wood.
- 100 sq. ft. of 1/2 by 12 in. soft wood.
- 1 piece 34 in. wide and 54 in. long hardwood.

- 30 pieces 1 by 1 in. 48 in. long.

[62]

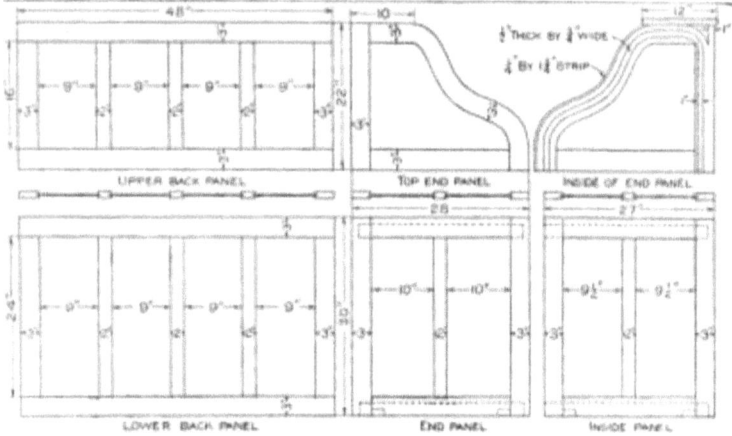

Rolltop Details

[63]

The upper and lower back panels are constructed very similar, the only difference being in the height. The inside edge of the 3-in. pieces is plowed with a 1/4-in. plow 3/8 in. deep exactly in the center and also both edges of each 2-in. piece. The 16-in. pieces in the upper back panel and the 24-in. pieces in the lower back panel must be cut 1/2 in. longer and a 1/4-in. tongue made on each end to fit into the plowed groove and form a mortise joint.

The upper back panel is filled in with four boards 9-1/2 in. wide and 16-1/2 in. long, while the four boards in the lower back panel are 9-1/2 in. wide and 24-1/2 in. long cut from the 1/4-in. hard wood. When the grooves are cut properly, the joints made perfect and the boards fitted to the right size, these two panels can be assembled and pressed together in cabinet clamps. This will make the outside dimensions as given in the drawing.

The end panels are made very similar to the lower back panel, the only difference being in the width of the filling boards, which are 10-1/2 in. for the outside end panels and 10 in. for the inside panels. One end panel and one inside panel make the sides of one pedestal.

As the end panels are 1 in. wider than the inside panels they overlap the back panel and cover up the rough ends of the boards. A 1-in. piece 2 in. wide is fastened at the top and bottom of each end and inside panels as shown by the dotted lines. The lower back panel is fastened on by turning screws through the back and into the ends of these pieces. The bottom pieces have 2-in. notches cut out, as shown, into which to fit two crosspieces across the bottom of the pedestal for holding the casters. The top end panels are made as shown in [64] the drawing, the inside edge of the pieces being plowed out, making a groove the same size as in the other pieces of the panels. The panel board is cut to the proper shape from the 1/4- by 24-in. material. The length given in the material list will be sufficient if the pointed ends are allowed to pass each other when laying out the design.

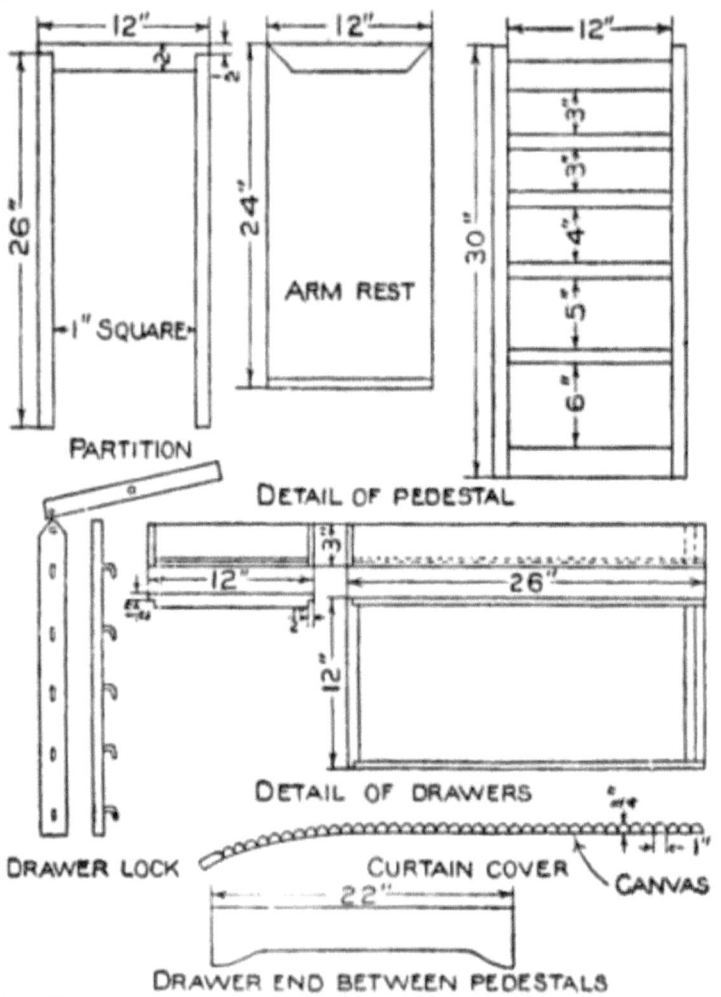

Details

Instead of cutting a groove for the roll top curtain, one is made by fastening a 1/2-by 3/4-in. strip 7/8 in. down from the edge and on the inside of the panel. A thin 1/4-by 1-3/4-in. strip is bent to form the shape of the edge and fastened with round-headed brass screws. A 1-in. piece is fastened at the back and a groove cut into it as shown by the dotted line into which to slide a 1/4-in. back board. The top is a 12-in. board 54 in. long.

As both pedestals are made alike, the detail of [65] only one is shown. The partitions upon which the drawers slide are made up from 1-in. square material with a 2-in. end fitted as shown. Dimensions are given for the divisions of each drawer, but these can be changed to suit the builder. The detail of one drawer is shown, giving the length and width, the height being that of the top drawer. The roll top curtain is made up from 1-in. pieces 3/4 in. thick and 48 in. long, cut in an oval shape on the outside, tacked and glued to a piece of strong canvas on the inside. The end piece is 2 in. wide, into which two lift holes or grooves are cut and a lock attached in the middle of the edge. A drawer lock can be made as shown and attached to the back panel and operated by the back end of the roll top curtain when it is opened and closed.

The top board, which is 34 by 54 in., can be fitted with end pieces as shown or left in one piece with the edges made rounding.

At this point in the construction of the parts they can be put together. The sides of each pedestal are fastened together by screws passed through the 1-in. square pieces forming the partition and into the sides of the panels. When each pedestal is put together the lower back panel is fastened to them with screws turned into the pieces provided as stated in making the end panels. The top board is now adjusted with equal edges projecting and fastened in position with finishing nails. As the top panels cover directly over where the nails are driven, the heads will not show. The upper back panel is fastened to the curved ends and the whole top held to the top board with cast corner brackets that can be purchased at any hardware store. The top [66] should not be drawn together too close before the 1/4-in. back board is put in the grooves and the roll top curtain placed in position.

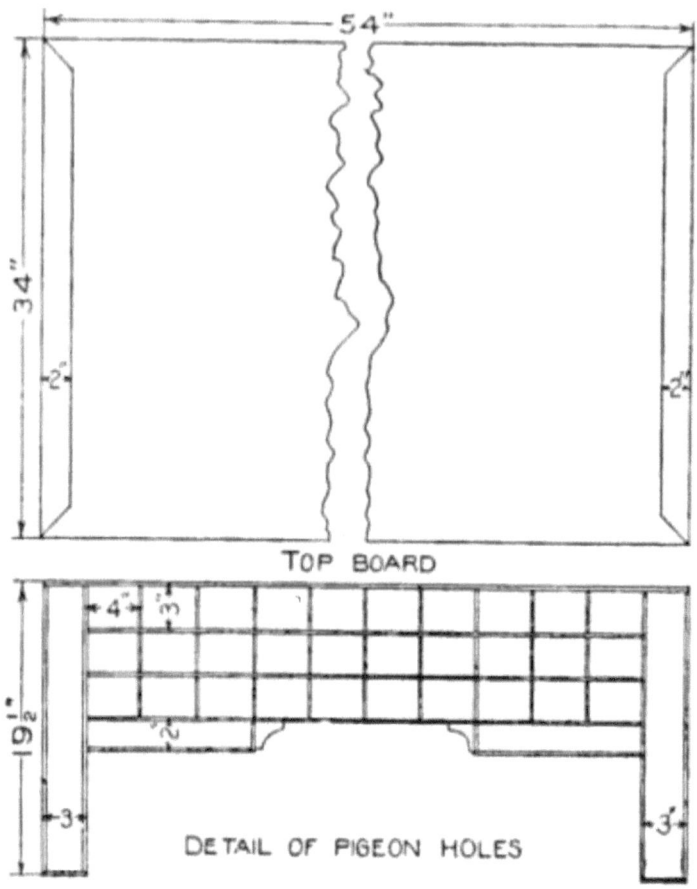

Detail of Pigeonholes

The detail showing the pigeon holes gives sizes for 30 openings 3 by 4 in., two book stalls at the ends, 3 in. wide, and two small drawers. This frame is built up as shown from the 3/8-in. soft wood, and fastened in the back part of the top with small brads.

[67]

HOW TO MAKE A ROMAN CHAIR

In making this roman chair, as well as other articles of mission furniture, the materials can be ordered from the mill with much of the hard work completed. Order the stock to make this chair as follows:

- 4 posts, 1-7/8 by 1-7/8 by 30 in.
- 2 top rails, 7/8 by 2-3/4 by 20 in.
- 2 bottom rails, 7/8 by 2-1/4 by 20 in.
- 2 rails, 7/8 by 4 by 16 in.
- 2 side rails, 7/8 by 4 by 28 in.
- 1 stretcher, 7/8 by 3 by 30 in.

The Roman Chair

Have all these pieces mill planed on the four sides straight and square, also have them sandpapered on the four sides of each. Plain sawed white [68] or red oak finishes nicely and is easily obtained. The sizes are specified exact as to thickness and width, but the lengths are longer than is needed. This is to allow for cutting and fitting.

Begin by squaring one end of each post; measure the length 28 in. and, placing all of them side by side, square a line across the four, saw, then plane these ends square. The top and bottom side rails are treated in a similar manner, their length being 19-1/8 in. each. These pieces extend right through the posts projecting 5/8 in. beyond the surface. The mortises in the posts must be cut smoothly and of exact size. Wood pins fasten these rails and posts together. The other rails have tenons 1/2 by 3 in. shouldered on the two edges and one side. The mortise in the post is placed central. On the ends of the chair the shouldered side is turned in (see photograph), while on the front and back they are turned out. Miter the ends of these tenons. These tenons are to be glued and clamped—the ends of the chair being put together first. When this is dry the sides are clamped. The stretcher should have its ends shouldered on the two edges so as to make a 2-1/2-in. tenon. Allow the tenons to extend 1-1/8 in. beyond the cross rail and cut mortises in these tenons for the keys.

All projecting tenons, as well as the tops and bottoms of the posts, should be chamfered about 1/8 in. For the seat, screw cleats to the insides of the rails and place a platform of thin boards so that its top surface is 1/2 in. below the top of the rails.

A cushion can be made, as shown in the photograph, by lacing with leather thongs two pieces of Spanish leather cut to proper length and width. [69] When nearly laced fill with any of the common upholsterer's fillings.

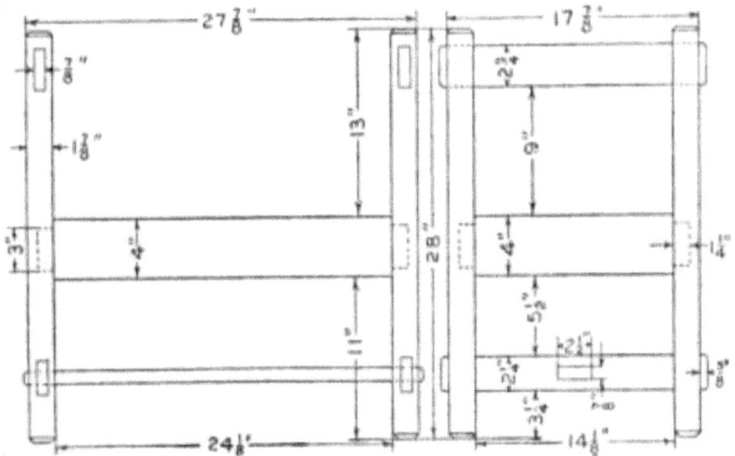

Details of Parts of Chair

For a brown stain, dissolve by boiling in 4 oz. of water, extract of logwood the size of a walnut. Apply hot and repeat until the desired color is obtained. Stains can be bought ready prepared, however, and are quite satisfactory. Finish by applying several coats of wax.

HOW TO MAKE A SETTEE

This handsome piece of mission furniture is designed to be made up in three different pieces as desired, the only changes necessary being in the length of the one front and the two back rails. The settee can be made into a three-cushion length by adding the length of another cushion to the dimensions of the one front and two back rails. A companion piece chair can be made by using suitable length rails to admit only one cushion. The following stock list of materials ordered mill-planed and sandpapered will be sufficient to make up the settee as illustrated. Oak is the most suitable wood which can be finished in either mission or a dark golden oak.

- 3 rails 1 by 4 by 52-1/4 in.
- 4 end rails 1 by 4 by 24-1/4 in.
- 4 posts 2-1/4 in. square by 34-1/2 in.
- 13 slats 1/2 by 5 by 21-1/4 in.
- 2 cleats 1 in. square by 51 in.

All the rails are mortised into the posts for a depth of 5/8 in., also the slats are mortised 5/8 in. into the rails. The material list gives the exact dimensions for the rails and slats as they will not need to be squared for entering the mortises, provided you are careful to get all lengths cut to dimensions. When cutting the mortises take care to get them square and clean. The posts have 1/2 in. extra added for squaring up and cutting the corners sloping on the top ends.

The joints are all put together with glue. Nails can be driven into the posts intersecting the tenons of the rails on the inside, as they will not show and will help to make the settee more solid.

[71]

A Complete Two-Cushion Settee

[72]

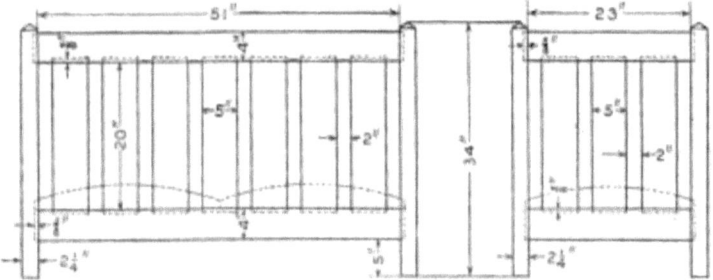

Details of a Mission Settee

The cushions can be made with or without springs as desired. If made without springs, 15 slats must be provided in the material list 1/2 in. thick, 2 in. wide and 24 in. long to be placed on the cleats fastened to the inside of each bottom rail. The two cleats are fastened one on each inside of the front and back rails with screws. The location as to height of these cleats will depend upon the kind of cushions used. The parts necessary to make the cushions with springs are as follows:

- 4 pieces 1 by 2-1/2 by 26 in.

- 8 pieces 1 by 2-1/2 by 24 in.
- 4 pieces 1 by 2-1/2 by 22 in.
- 32 8-in. springs.
- 2 pieces leather about 29 by 31 in.

[73]

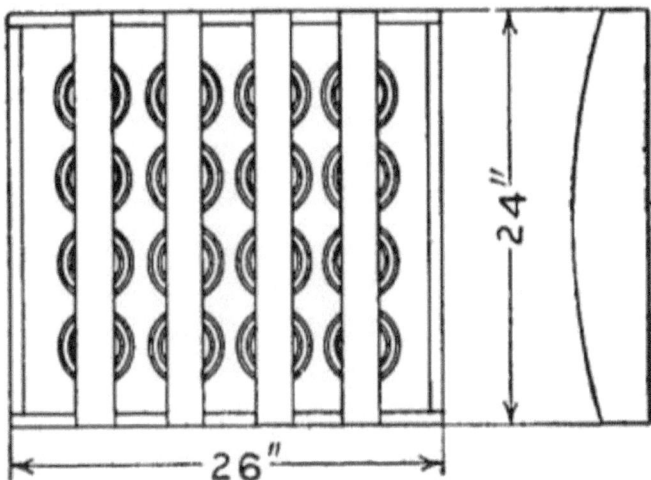

Details of the Cushion

An open box is made from two 26-in. and two 22-in. pieces, and across the bottom are mortised and set in four 24-in. pieces to form slats on which to set the springs. The tops of the springs are tied or anchored with stout cords running in both directions and fastened to the inside of the pieces forming the open box. These should be tied in such manner as to hold each spring so it cannot slip over and come in contact with another spring.

Roan or pebbled leather are very popular for cushions for this style of furniture. The leather is drawn over the springs and tacked to the outside of the open box frame. When complete the cushions are set in loose on the cleats, which should, in this case, be placed about 1 in. from the top of the rails.

[74]

HOW TO MAKE A PYROGRAPHER'S TABLE

Fig. 1   Fig. 2
Convenient Pyrographer's Table

Any pyrographer will appreciate the construction of the table and cabinet as illustrated. Anyone doing burnt wood work will know the annoyance of building up a steady support for the arm to the level of the article on which the work is to be done. The size of this table may be made to suit the surroundings and the space of the builder. Figure 1 shows the table with a slot cut in the side support in which to place the thumb screw of the bracket as shown on top of the table. It will be noticed, Fig. 2, that while both drawer and cabinet are available for storing the apparatus, they are not in the way of the operator [75] while sitting at his work; the drawer overhangs the knees and the cabinet is far enough back not to interfere with sitting up close to the work. The bracket shelf slides in the slot at the side of the table, and is fastened to any height by the thumb screw There is also a smaller slide bracket on the shelf to clamp irregular objects to the side of the table. The thumb screws, hinges and drawer pulls can be purchased from any hardware store. When the table is not in use for pyrography it can be used for a writing table or a round top provided and attached on which to play games. When used for this purpose the bracket, as well as the pyrographic outfit, is stowed away in the cabinet as shown in Fig. 3.

Storage for Apparatus

MISSION STAINS

What is mission oak stain? There are many on the market, with hardly two alike in tone. The true mission oak stain may be said to show a dull gray, the flakes showing a reddish tint, while the grain of the wood will be almost a dead black. To produce such a stain take 1 lb. of drop black in oil and 1/2 oz, of rose pink in oil, adding a gill of best japan drier, thinning with three half-pints of turpentine. This will make about 1 qt. of stain. Use these proportions for a larger quantity of stain. Strain it through cheese cloth. Japan colors will give a quicker drying stain than that made with oil colors, and in this case omit the japan and add a little varnish to bind it.

One of the most popular of all the fancy oaks has been that known as Flemish, and this in spite of its very somber color, says Wood Craft. There are several ways of producing Flemish finish; you can fill the wood with a paste filler strained with raw umber, and when dry apply a stain of transparent flat raw umber, and for the darker shades of finish use drop black with the umber. Varnish and rub down.

According to a foreign technical journal, French workmen mahoganize various kinds of woods by the following method: The surface of the wood to be stained is made perfectly smooth. Then it is given a coating of dilute nitric acid which is rubbed well into the wood fiber. Then it is stained with a mixture made by dissolving 1-1/2 oz. of dragon's blood in a pint of alcohol, this solution being filtered, and then there is added to it one-third of its weight of sodium carbonate. Apply this mixture with a brush, and repeat the coats at intervals until the [77] surface has the appearance of polished mahogany. In case the luster should fail it may be restored by rubbing with a little raw linseed oil. The description of the process is meager, and hence he who would try it will have to experiment a little.

A good cheap mission effect for oak is to mix together equal parts of boiled linseed oil and good asphaltum varnish, and apply this to the wood with a brush; in a minute or so you may rub off surplus with a rag, and when dry give a coat of varnish. A gallon of this stain will cover about 600 sq. ft.

FILLING OAK

A very good hardwood filler for oak, either for a natural or golden effect, may be made from two parts of turpentine and one part of raw linseed oil, with a small amount of good japan to dry in the usual time. To this liquid add bolted gilder's whiting to form a suitable paste, it may be made thin enough for use, if to be used at once, or into a stiff paste for future use, when it can be thinned down for use, says Woodworkers' Review. After applying a coat of filler, let stand until it turns gray, which requires about 20 minutes, depending upon the amount of japan in the filler, when it should be rubbed off with cotton waste or whatever you use for the purpose. A filler must be rubbed well into the wood, the surplus only being removed. The application of a coat of burnt umber stain to the wood before filling is in order, which will darken the wood to the proper depth if you rub off the surplus, showing the grain and giving a golden oak effect. The filling should stand at least a day and night before applying shellac and varnish.

[78]

WAX FINISHING

In wax-finishing hardwoods, use a paste filler and shellac varnish to get a good surface. Of course, the wax may also be rubbed into the unfilled wood but that gives you quite a different effect from the regular wax polish, says a correspondent of Wood Craft. With soft woods you first apply a stain, then apply a liquid filler or shellac, according to the quality of work to be done. The former for the cheaper job. The usual proportion of wax and turpentine is two parts of the former to one part of the latter, melting the wax first, then adding the spirits of turpentine. For reviving or polishing furniture you can add three or four times as much turpentine as wax, all these proportions to be by weight. To produce the desired eggshell gloss, rub vigorously with a brush of stiff bristles or woolen rag.

THE FUMING OF OAK

Darkened oak always has a better appearance when fumed with ammonia. This process is rather a difficult one, as it requires an airtight case, but the description herewith given may be entered into with as large a case as the builder cares to construct.

Oak articles can be treated in a case made from a tin biscuit box, or any other metal receptacle of good proportions, provided it is airtight. The oak to be fumed is arranged in the box so the fumes will entirely surround the piece; the article may be propped up with small sticks, or suspended by a string. The chief point is to see that no part of the wood is covered up and that all surfaces are exposed to the fumes. A saucer of ammonia is placed [79] in the bottom of the box, the lid or cover closed, and all joints sealed up by pasting heavy brown paper over them. Any leakage will be detected if the nose is placed near the tin and farther application of the paper will stop the holes. A hole may be cut in the cover and a piece of glass fitted in, taking care to have all the edges closed. The process may be watched through the glass and the article removed when the oak is fumed to the desired shade. Wood stained in this manner should not be French polished or varnished, but waxed.

The process of waxing is simple: Cut some bees-wax into fine shreds and place them in a small pot or jar. Pour in a little turpentine, and set aside for half a day, giving it an occasional stir. The wax must be thoroughly dissolved and then more turpentine added until the preparation has the consistency of a thick cream. This can be applied to the wood with a rag and afterward brushed up with a stiff brush.

HOW TO MAKE BLACK WAX

When putting a wax finish on oak or any open-grained wood, the wax will often show white streaks in the pores of the wood. These streaks cannot be removed by rubbing or brushing. Prepared black wax can be purchased, but if you do not have any on hand, ordinary floor wax can be colored black. Melt the floor wax in a can placed in a bucket of hot water. When the wax has become liquid mix thoroughly into it a little drop black or lampblack. Allow the wax to cool and harden. This wax will not streak, but will give a smooth, glossy finish.

[80]

THE 40 STYLES OF CHAIRS

There are 40 distinct styles of chairs embracing the period from 3000 B.C. to 1900 A.D.—nearly 7,000 years. Of all the millions of chairs made during the centuries, each one can be classified under one or more of the 40 general styles shown in the chart. This chart was compiled by the editor of Decorative Furniture. The Colonial does not appear on the chart because it classifies under the Jacobean and other styles. A condensed key to the chart follows:

Egyptian.—3000 B.C. to 500 B.C. Seems to have been derived largely from the Early Asian. It influenced Assyrian and Greek decorations, and was used as a motif in some French Empire decoration. Not used in its entirety except for lodge rooms, etc.

Grecian.—700 B.C. to 200 B.C. Influenced by Egyptian and Assyrian styles. It had a progressive growth through the Doric, Ionic and Corinthian periods. It influenced the Roman style and the Pompeian, and all the Renaissance styles, and all styles following the Renaissance, and is still the most important factor in decorations today.

Roman.—750 B.C. to 450 A.D. Rome took her art entirely from Greece, and the Roman is purely a Greek development. The Roman style "revived" in the Renaissance, and in this way is still a prominent factor in modern decoration.

Pompeian.—100 B.C. to 79 A.D. Sometimes called the Grecian-Roman style, which well describes its components. The style we know as Greek was the Greek as used in public structures. The Pompeian is our best idea of Greek domestic decoration. Pompeii was long buried, but when rediscovered it promptly influenced all European styles, including Louis XVI, and the various Georgian styles.

Byzantine.—300 A.D. to 1450 A.D. The "Eastern Roman" style, originating in the removal of the capital of the Roman Empire to Constantinople (then called Byzantium). It is a combination of Persian and Roman. It influenced the various Moorish, Sacracenic and other Mohammedan styles.

Gothic.—1100 to 1550. It had nothing to do with the Goths, but was a local European outgrowth of the Romanesque. It spread all over Europe, and reached its climax of development about 1550. It was on the Gothic construction that the Northern European and English Renaissance styles were grafted to form such styles as the Elizabethan, etc.

[81]

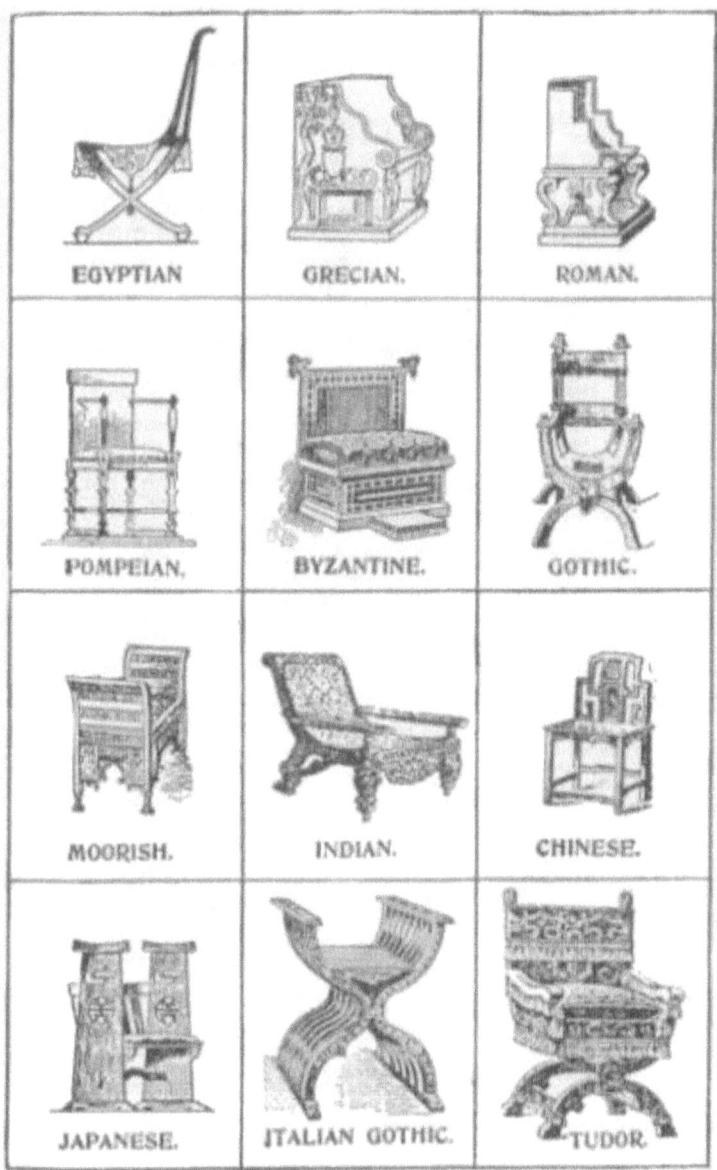

Chairs 1

[82]

Moorish.—700 to 1600. The various Mohammedan styles can all be traced to the ancient Persian through the Byzantine. The Moorish or Moresque was the form taken by the Mohammedans in Spain.

Indian.—2000 B.C. to 1906 A.D. The East Indian style is almost composite, as expected of one with a growth of nearly 4,000 years. It has been influenced repeatedly by outside forces and various religious invasions, and has, in turn, influenced other far Eastern styles.

Chinese.—3500 B.C. to 1906 A.D. Another of the ancient styles. It had a continuous growth up to 230 B.C., since when it has not changed much. It has influenced Western styles, as in the Chippendale, Queen Anne, etc.

Japanese.—1200 B.C. to 1906 A.D. A style probably springing originally from China, but now absolutely distinct. It has influenced recent art in Europe and America, especially the "New Art" styles.

Italian Gothic.—1100 to 1500. The Italian Gothic differs from the European and English Gothic in clinging more closely to the Romanesque-Byzantine originals.

Tudor.—1485 to 1558. The earliest entry of the Renaissance into England. An application of Renaissance to the Gothic foundations. Its growth was into the Elizabethan.

Italian Renaissance, Fifteenth Century.—1400 to 1500. The birth century of the Renaissance. A seeking for revival of the old Roman and Greek decorative and constructive forms.

Italian Renaissance, Sixteenth Century.—1500 to 1600. A period of greater elaboration of detail and more freedom from actual Greek and Roman models.

Italian Renaissance, Seventeenth Century.—1600 to 1700. The period of great elaboration and beginning of reckless ornamentation.

Spanish Renaissance.—1500 to 1700. A variation of the Renaissance spirit caused by the combination of three distinct styles—the Renaissance as known in Italy, the Gothic and the Moorish. In furniture the Spanish Renaissance is almost identical with the Flemish, which it influenced.

Dutch Renaissance.—1500 to 1700. A style influenced alternately by the French and the Spanish. This style and the Flemish had a strong influence on the English William and Mary and Queen Anne styles, and especially on the Jacobean.

German Renaissance.—1550 to 1700. A style introduced by Germans who had gone to Italy to study. It was a heavy treatment of the Renaissance spirit, and merged into the German Baroque about 1700.

Francis I.—1515 to 1549. The introductory period when the Italian Renaissance found foothold in France. It is almost purely Italian, and was the forerunner of the Henri II.

Henri II.—1549 to 1610. In this the French Renaissance became differentiated from the Italian, assuming traits that were specifically French and that were emphasized in the next period.

Louis XIII.—1616 to 1643. A typically French style, in which but few traces of its derivation from the Italian remained. It was followed by the Louis XIV.

Elizabethan.—1558 to 1603. A compound style containing traces of the Gothic, much of the Tudor, some Dutch, Flemish and a little Italian. Especially noted for its fine wood carving.

Jacobean.—1603 to 1689. The English period immediately following the Elizabethan, and in most respects quite similar. The Dutch influence was, however, more prominent. The Cromwellian, which is included in this period, was identical with it.

William and Mary.—1689 to 1702. More Dutch influences. All furniture lighter and better suited to domestic purposes.

[83]

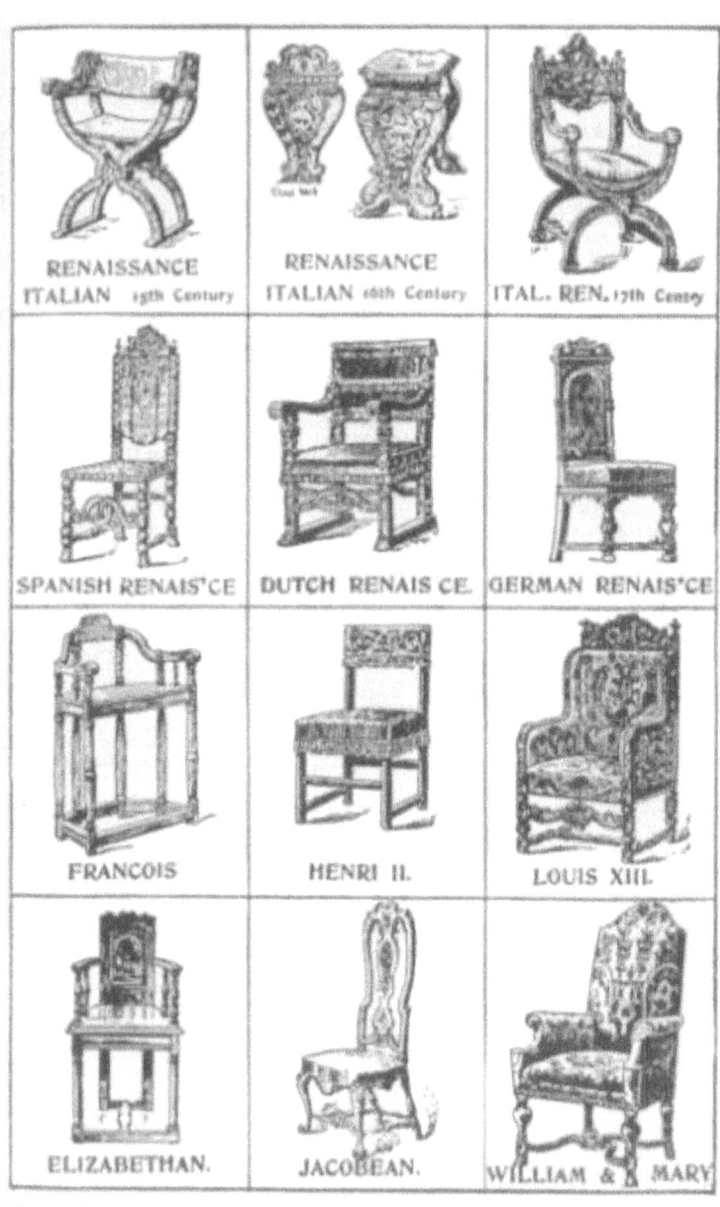

Chairs 2

[84]

Queen Anne.—1702 to 1714. Increasing Dutch influences. Jacobean influence finally discarded. Chinese influence largely present.

Louis XIV.—1643 to 1715. The greatest French style. An entirely French creation, marked by elegance and dignity. Toward the end of the period it softened into the early Rococo.

Georgian.—1714 to 1820. A direct outgrowth of the Queen Anne, tempered by the prevailing French styles. It includes Chippendale, Hepplewhite and Sheraton, but these three great cabinetmakers were sufficiently distinct from the average Georgian to be worthy separate classification.

Chippendale.—1754 to 1800. The greatest English cabinet style. Based on the Queen Anne, but drawing largely from the Rococo, Chinese and Gothic, he produced three distinct types, viz.: French Chippendale, Chinese Chippendale and Gothic Chippendale. The last is a negligible quantity.

Louis XV.—1715 to 1774. The Rococo period. The result of the efforts of French designers to enliven the Louis XIV, and to evolve a new style out of one that had reached its logical climax.

Hepplewhite.—1775 to 1800. Succeeded Chippendale as the popular English cabinetmaker. By many he is considered his superior. His work is notable for a charming delicacy of line and design.

Louis XVI.—1774 to 1793. The French style based on a revival of Greek forms, and influenced by the discovery of the ruins of Pompeii.

Sheraton.—1775 to 1800. A fellow cabinetmaker, working at same time as Hepplewhite. One of the Colonial styles (Georgian).

R. & J. Adam.—1762 to 1800. Fathers of an English classic revival. Much like the French Louis XVI and Empire styles in many respects.

Empire.—1804 to 1814. The style created during the Empire of Napoleon I. Derived from classic Roman suggestions, with some Greek and Egyptian influences.

New Arts.—1900 to date. These are various worthy attempts by the designers of various nations to create a new style. Some of the results are good, and they are apt to be like the "little girl who had a little curl that hung in the middle of her forehead," in that "when they are good they are very, very good, but when they are bad they are horrid." [85]

Chairs 3

[86]

Chairs 4
[87]

HOW TO MAKE A PIANO BENCH

Piano Bench

All the material used in the making of this piano bench is 1 in. thick, excepting the two rails, which are 7/8 in. thick. The bench can be made from any of the furniture woods, but the case may demand one made from mahogany. If so, this wood can be purchased from a piano factory. The following stock list of materials may be ordered from a mill, planed and sandpapered:

- 1 top, 1 by 16 by 36-1/2 in.
- 2 ends, 1 by 14 by 18 in.
- 1 stretcher, 1 by 4 by 31-1/2 in.
- 2 side rails, 7/8 by 4 by 29-1/2 in.
- 2 keys, 1 by 1 by 3-1/2 in.
- 6 cleats, 1 by 1 by 4 in.

The dimensions given, with the exception of the keys and cleats, are 1/2 in. longer than necessary for squaring up the ends.

The two rails are cut slanting from a point 1-1/2 [88] in. from each end to the center, making them only 3 in. wide in the middle. The rails are "let into" the edges of the ends so the outside of the rails and end boards will be flush. The joints are put together with glue and screws. The cleats are fastened with screws to the inside of the rails and to the top. The stretcher has a tenon cut on each end which fits into a mortise cut in each end. The tenons will have sufficient length to cut the small mortise for the key.

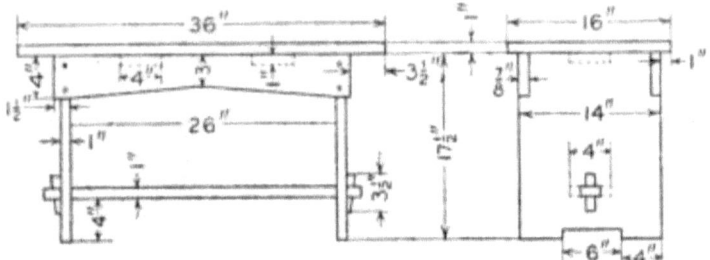

Piano Bench Details

The kind of wood used will determine the color of the stain for the finish. This also depends on matching other pieces of furniture.

[89]

HOW TO MAKE A MISSION SHAVING STAND

This attractive and useful piece of mission furniture will be appreciated by the person that does his own shaving. The shaving stand can be made at home by a handy man in his spare time as the stock can be ordered from a mill ready for making the joints and attaching the few pieces of hardware. The following is a stock list of materials:

- 4 posts 1-1/2 in. square by 50-1/2 in.
- 4 slats 7/8 by 1 by 32-1/2 in.
- 2 cross rails 1 by 1-1/2 by 15 in.
- 2 end rails 1 by 1-1/2 by 13 in.
- 1 top 7/8 by 16-1/2 by 19-1/2 in.
- 1 bottom 7/8 by 15 by 17 in.
- 2 ends 7/8 by 12-1/2 in. square.
- 1 back 7/8 by 12-1/2 by 14-1/2 in.
- 1 door 7/8 by 6-1/2 by 12-1/2 in.
- 2 drawer ends 7/8 by 5-1/2 by 7-1/2 in.
- 1 partition 7/8 by 12 by 14 in.
- 1 partition 7/8 by 7 by 14 in.
- 7 pieces of soft wood 1/2 by 7-1/2 by 12 in.
- 2 posts 1 in. square by 10-1/2 in.
- 1 bottom piece 7/8 by 1-1/2 by 18-1/2 in.
- 4 mirror frame pieces 7/8 by 1-1/2 by 14-1/2 in.
- 2 sticks for pins.
- 2 hinges
- 1 lock
- 2 drawer pulls
- 1 beveled glass mirror 11-1/2 by 11-1/2 in.

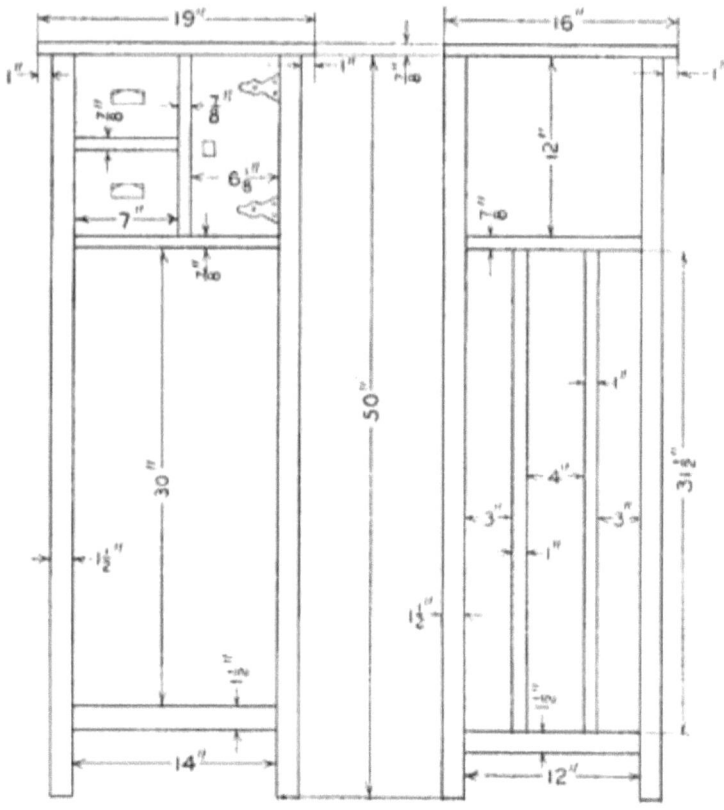

Shaving Stand Details

While this piece of furniture can be made in any kind of wood, the novice will find that quarter-sawed oak will work up and finish better than the other woods. The stock list given has dimensions 1/2 in. larger in some instances for dressing and squaring where necessary.

The tenons and mortises are first cut for the crosspieces at the bottom of the posts, and, as it is [90] best to use dowels at the top, holes are bored in the bottom piece and also the ends of the slats for pins. The bottom piece is also fastened to the posts with dowels. The bottom must have a square piece cut out from each corner almost the same size as the posts. When setting the sides together the end board and posts can be doweled and glued together [91] and after drying well the posts can be spread apart far enough to insert the bottom rail and two slats. The rail and slats should be tried for a bit before putting on any glue, which may save some trouble.

[92]

Shaving Stand

Complete

After the sides are put together, the back is put in and glued. The top is then put on and fastened with cleats from the inside. The partitions are put in as shown and the door fitted. Two drawers are made from the ends and the soft wood material. The drawer ends may be supplied with wood pulls of the same material or matched with metal the same as used for the hinges.

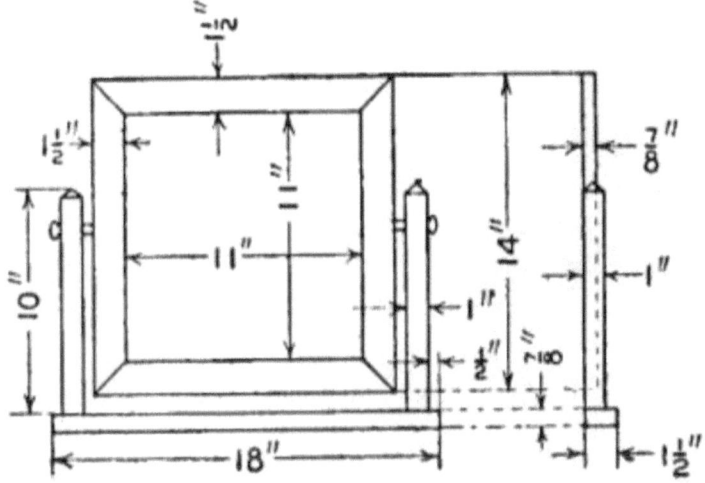

Mirror Frame and Standards Details

The pieces for the mirror frame must be rabbeted 1/2 in. deep to take the glass, and the ends joined together with a miter at each corner. The two short posts are tenoned and mortises cut in the bottom piece for joints and these joints well glued together. The bottom piece is then fastened to the top board of the stand. This will form the standards in which to swing the mirror and its frame. This is done with two pins inserted in holes bored through the standards and into the mirror frame.

After the parts are all put together, cleaned and sandpapered, the stand is ready for the finish.

[93]

A MISSION WASTE-PAPER BASKET

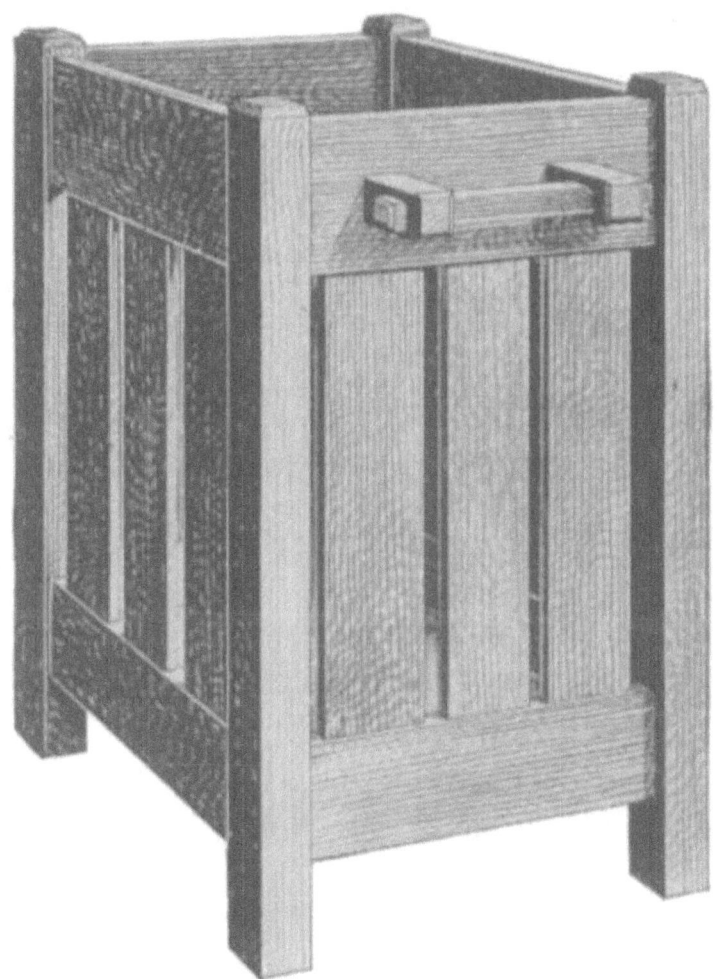

Waste-Paper Basket to Match Library Table

The basket shown in the accompanying sketch is designed to be used with a library table having slats in the ends and wooden handles on the drawers. The finish is made to match that of the table by fuming, when completely assembled, in a large-size [94] size, clean garbage can, with fumes of concentrated ammonia.

[95]

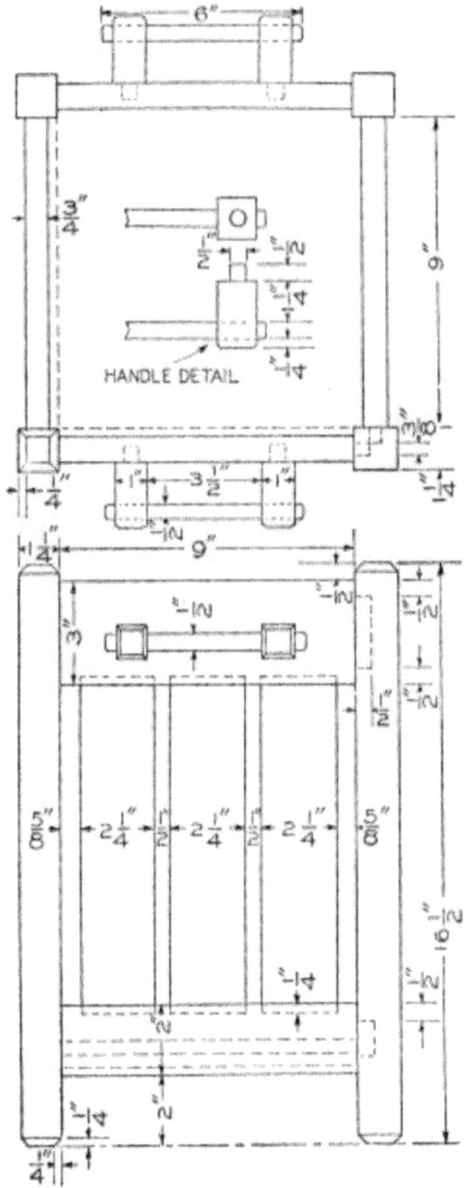

Detail of Waste-Paper

Basket

The following quarter-sawed white-oak stock should be procured in the exact dimensions given. This may be had, planed and cut to lengths, from a mill for a slight extra charge. It is advisable not to have them sandpapered, as the very coarse sandpaper generally used, gives a bad surface for finishing.

- 4 posts, 1-1/4 by 1-1/4 by 16-1/2 in., S-4-S.
- 4 rails, 3/4 by 3 by 10-1/4 in., S-2-S.
- 4 rails, 3/4 by 2 by 10-1/4 in., S-2-S.
- 12 slats, 3/8 by 2-1/4 by 9-1/2 in., S-2-S.
- 4 handle pieces, 1 by 1 by 2-1/2 in., S-4-S.
- 2 handle pieces, 1/2 by 1/2 by 6 in., S-4-S.
- 1 bottom, 3/8 by 9-1/2 by 9-1/2 in., S-2-S.

See that the posts are absolutely square cross section. Mark with a pencil — not gauge — the chamfers on the ends of the posts and plane them off.

Carefully mark the tenons on the ends of all the rails with a knife and gauge lines. Be sure that the distance from the tenon shoulder at one end of rail to the shoulder at the other end is exactly the same on each rail. Cut the tenons, using a backsaw and chisel.

Arrange the pieces as they are to stand in the finished basket, and number each tenon and mortise. Mark all the mortises on the posts, being sure to keep the distances between the top and lower rail the same on each post. Cut each mortise to fit the correspondingly numbered tenon. Next, mark the mortises for the slats in the rails, allowing the whole slat to go in 1/4 in.

The handles are next in order. The pieces going into the rail should be fastened with a round [96] 1/2-in. tenon cut on one end and glued in place. The crosspiece should be mortised all the way through these pieces and held in place by a brad from the under side.

Now put the whole basket together without gluing, in order that errors, if any, may be detected.

If everything fits perfectly, the basket is ready to be glued. For best results hot glue should be used. First glue up two opposite sides with the slats in place. Clamps must be used. When these have set for at least 24 hours, the other rails and slats may be glued in place and clamped. It is a good idea to pin the tenons in place with two 1-in. brads driven from the inside.

The handles are then glued in place, using hand screws to hold them until the glue sets. The bottom should rest on thin cleats, without being nailed to them, so that it may be removed when the basket is to be emptied of small papers, etc.

Before applying the stain, see that all glue spots are removed and all surfaces sanded to perfect smoothness. If a fumed finish is not desired, any good stain may be used, after which a thin coat of shellac and two coats of wax should be applied. Allow plenty of time for drying between the coats.

A CELLARETTE PEDESTAL

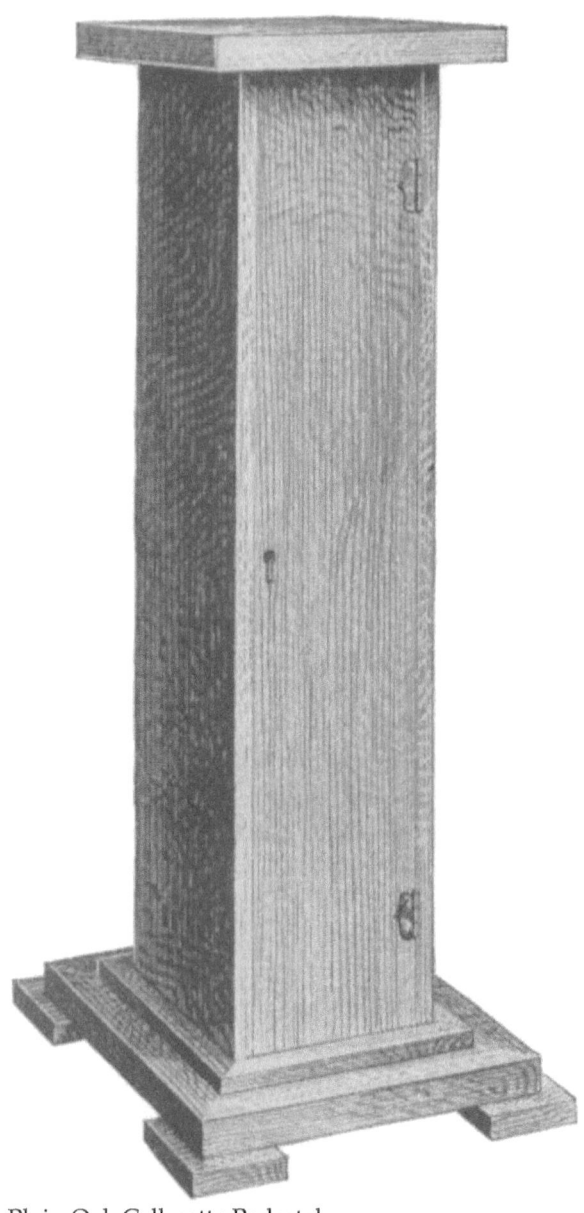

Plain-Oak Cellarette Pedestal

The illustration shows a unique article for the den. It serves as a pedestal and has one side which opens on hinges allowing the inside to be used as a smoker's cabinet or cellarette. All the lines are straight and the corners square, making it easy to [97] construct. White oak will make up best, although ash, birch or southern pine may be used with good effect.

[98] Stock of the following sizes should be bought, surfaced and cut to width and length:

- 2 top pieces, 7/8 by 12 by 12 in., S-2-S.
- 2 base pieces, 7/8 by 14 by 14 in., S-2-S.
- 2 sides, 7/8 by 8 by 35-5/8 in., S-2-S.
- 1 back, 7/8 by 6-1/4 by 35-5/8 in., S-2-S.
- 1 door, 7/8 by 6-1/4 by 34-3/4 in., S-2-S.
- 4 blocks, 7/8 by 4 by 4 in., S-2-S.
- 4 shelves, 7/8 by 6-1/4 by 6-1/4 in., S-2-S.
- 4 pieces, 7/8 by 1 by 10 in., S-4-S.

Make the top and base of two pieces, glued and screwed together with the grain crossed. This method prevents warping. To keep the end grain from showing, a strip of 3/8-in. lumber may be put on all around as shown in the drawing.

Have the sides, front and back squared up perfectly. The sides are to overlap the back and to be fastened to it with round-head brass or blue screws. To the center of the top and base attach one of the 6-1/4-in. square pieces. Over these, fit the sides and back and fasten them with screws or nails. The four corner blocks are now put under the base.

Two or more shelves may be set in as shown. Brass or copper hinges will look well if a dark stain is to be used.

Around the sides and back a 1-in. strip should be fastened to the base to give added strength.

If a dull finish is desired, apply two coats of stain and two of prepared wax. If a polished surface is wanted, first fill the pores of the wood with any standard filler, which can be purchased at a paint store. After this has dried partly, rub off any surplus filler, rubbing across the grain of the wood. [99] When perfectly dry apply one coat of shellac and as many coats of varnish as desired, rubbing down each coat, except the last, with No. 00 sandpaper and pumice stone.

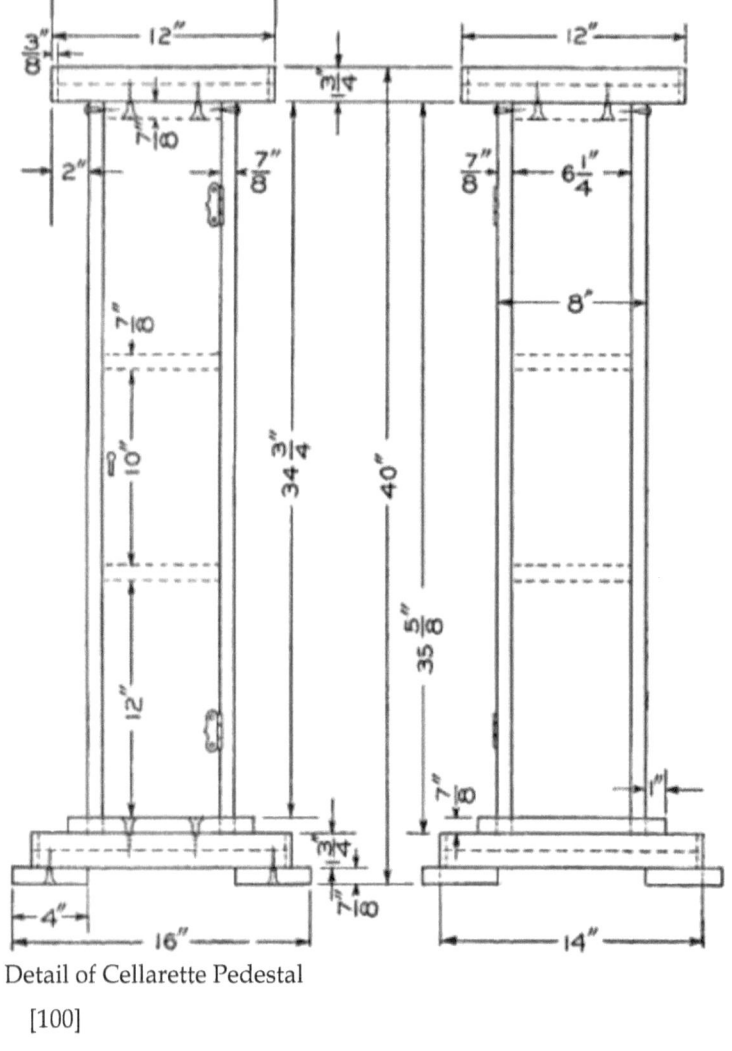

Detail of Cellarette Pedestal

[100]

A DRESSER

The dresser shown in the illustration was made of quarter-sawed white oak and finished golden and waxed. The mirror is of beveled glass and the following is the stock bill:

- 1 top, 3/4 by 19-1/2 by 33 in., S-2-S.
- 4 posts, 1-3/4 by 1-3/4 by 28 in., S-4-S.
- 4 end rails, 3/4 by 2-1/4 by 17 in., S-2-S.
- 4 stiles, 3/4 by 2-1/2 by 20 in., S-2-S.
- 2 panels, 3/16 by 12 by 18 in., S-2-S.
- 3 facings, 3/4 by 2-1/4 by 29 in., S-2-S.
- 2 top frame pieces, 3/4 by 2 by 32 in., S-2-S.
- 2 top frame pieces, 3/4 by 2 by 19 in., S-2-S.
- 2 mirror supports, 1 by 2 by 33 in., S-2-S.
- 1 mirror support, 3/4 by 2-1/2 by 33 in., S-2-S.
- 1 drawer front, 3/4 by 7-1/4 by 28 in., S-2-S.
- 1 drawer front, 3/4 by 6-1/4 by 28 in., S-2-S.
- 2 drawer fronts, 3/4 by 5-1/4 by 14 in., S-2-S.
- 1 partition, 3/4 by 1 by 6 in.
- 2 mirror-frame pieces, 3/4 by 2 by 40 in., S-2-S.
- 2 mirror-frame pieces, 3/4 by 2 by 20 in., S-2-S.

The following material list may be of common stock and not quarter-sawed:

- Mirror-backing pieces equivalent to 1/4 by 18-1/2 by 36 in., S-2-S.
- 2 cleats, 3/4 by 2 by 10 in., S-4-S.
- 4 drawer-support frame pieces, 3/4 by 2 by 29 in.
- 7 drawer-support frame pieces, 3/4 by 2 by 15 in.
- Slides taken from scrap stock, 3/4 by 1 by 15 in.
- 3 back pieces, 3/4 by 2-1/4 by 28 in., S-2-S.
- 2 back pieces, 1/4 by 8 by 28 in., S-2-S.
- 8 drawer sides, 1/2 by 7-1/4 by 17 in., S-2-S.

- 2 drawer backs, 3/8 by 7 by 27 in., S-2-S.
- 2 drawer backs, 3/8 by 7 by 13 in., S-2-S.
- 2 drawer bottoms, 3/8 by 15 by 27 in., S-2-S.
- 2 drawer bottoms, 3/8 by 15 by 13 in., S-2-S.

In working up the various parts proceed in the usual manner. If not thoroughly familiar with the various tool processes involved, it will be necessary to investigate pieces of near-by furniture and to read up some good text dealing with the processes involved. [101]

Dresser in

Quarter-Sawed Oak

[102]

The exact size of the mirror is 18 by 36 in. and the frame should be rabbeted to correspond.

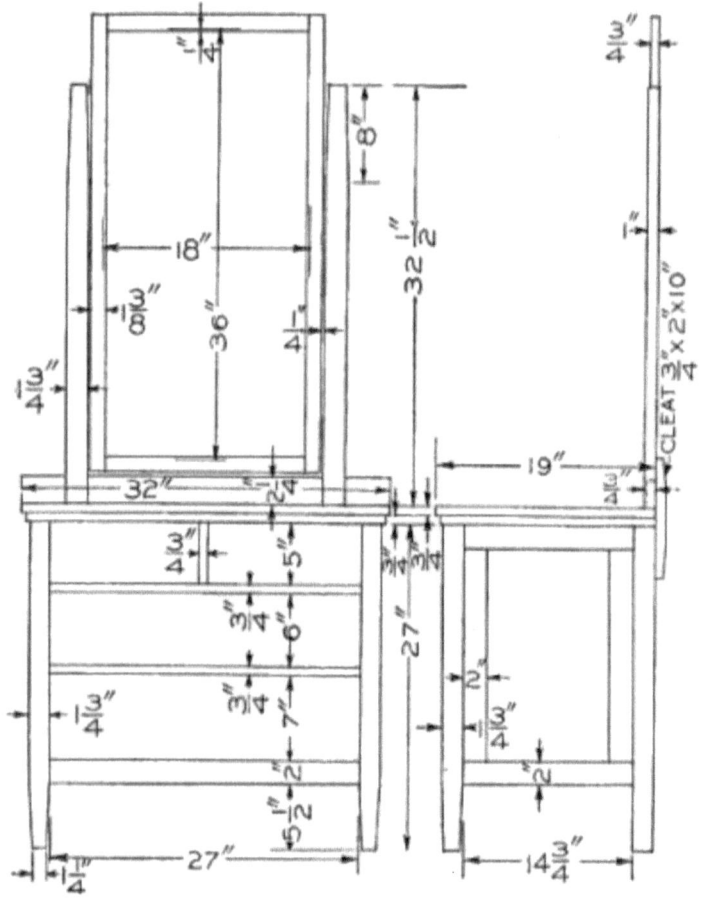

Detail of the Dresser

For a finish, a coat of paste filler colored so as to [103] give a rich golden brown should be applied first. Allow this to harden, after rubbing and polishing it in the usual manner, then apply a thin coat of shellac. Sand this lightly when hard, and over this apply a coat of orange shellac. Over the shellac put several coats of some good rubbing wax and polish each coat well. If a striking contrast is wanted for the medullary rays of the quartering, apply a golden-oak stain first. Sand this lightly, then apply a second coat diluted one-half with solvent and sand again lightly. Apply a thin coat of shellac, then, when dry, sand lightly and apply paste, and proceed as before.

A MISSION SIDEBOARD

Oak is the most suitable material for making this sideboard and it should be first-class stock, planed and cut to the dimensions given in the following list:

- 1 top, 7/8 by 22 by 48 in., S-2-S.
- 1 top shelf, 7/8 by 12 by 48 in., S-2-S.
- 1 bottom, 7/8 by 22 by 48 in., S-2-S.
- 2 back posts, 2 by 2 by 57 in., S-4-S.
- 2 front posts, 2 by 2 by 36 in., S-4-S.
- 2 standards, 2 by 2 by 20 in., S-4-S.
- 2 mirror rails, 7/8 by 2 by 47 in., S-2-S.
- 2 mirror rails, 7/8 by 2 by 20 in., S-2-S.
- 3 front and back rails, 7/8 by 3 by 46 in., S-2-S.
- 4 end rails, 7/8 by 3 by 20 in., S-2-S.
- 4 standard rails, 7/8 by 2 by 10 in., S-2-S.
- 2 vertical pieces, 7/8 by 19-1/2 by 22 in., S-2-S.
- 1 horizontal piece, 7/8 by 22 by 14-1/4 in., S-2-S.
- 1 drawer front, 7/8 by 6 by 14-1/4 in., S-2-S.
- 1 piece, 7/8 by 3 by 3 in.
- 4 vertical door pieces, 7/8 by 2 by 17 in., S-2-S.
- 4 horizontal door pieces, 7/8 by 2 by 15 in., S-2-S.
- 2 drawer sides, 7/8 by 5 by 14 in., S-2-S.
- 1 drawer bottom, 1/4 by 14 by 14-1/4 in., S-2-S.
- 1 back panel, 1/4 by 16-1/2 by 44-1/2 in., S-2-S.
- 2 door panels, 1/4 by 10-1/2 by 15-1/2 in., S-2-S.
- 2 side panels, 1/4 by 18-1/2 by 16-1/2 in., S-2-S.

[104]

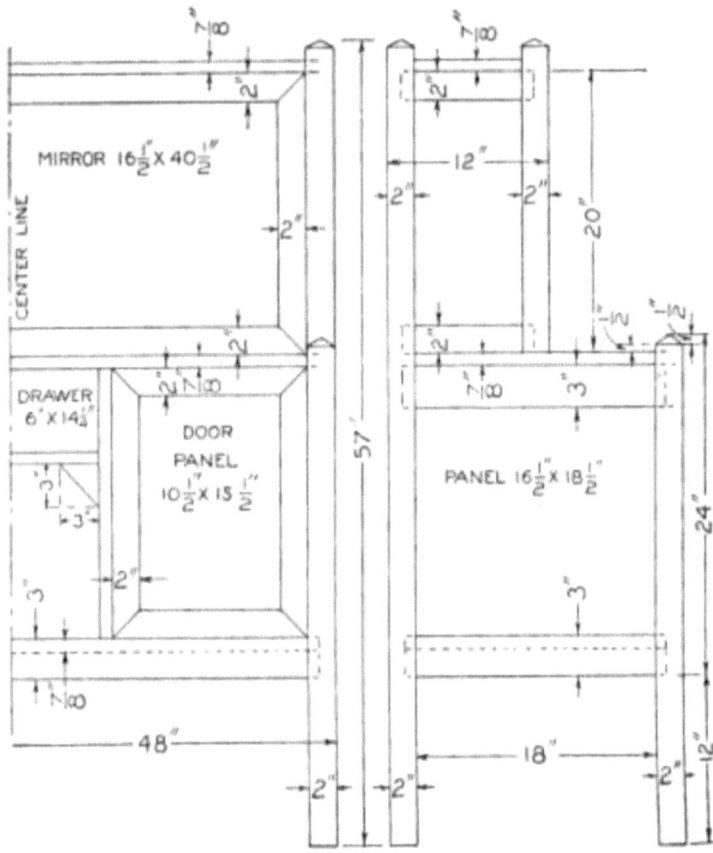

Detail of the Mission Sideboard

Begin work by cutting the posts to the length indicated in the detail drawing. The top ends are tapered with a 1/2-in. slant. These posts are cut in pairs and it is best to stand them up in the same [105] position they will be in the finished sideboard, and mark the sides to be mortised with a pencil. Also cut the grooves into which the panels are to fit. These are to be 1/4 in. wide and a little over 1/4 in. deep.

Mission Sideboard in Quarter-Sawed Oak

[106] The rails are cut with tenon ends to match the mortises, and also have grooves to receive the panels.

The bottom part of the back is closed with a panel and two rails, one at the same height from the floor as the front bottom rail, and the top one even with the under side of the top. The large panel is for the opening thus formed.

These parts are now put together, using plenty of good hot glue, and spreading it well on the mortises and tenon ends.

When drawing the frame together with the clamps, care must be taken to get it square.

After the glue is hard enough to remove the clamps, the top and bottom are put in place. The corners of the top are notched out to fit around the posts, while the bottom is cut to fit on the inside of the rails and is held in place by putting screws in at an angle through the bottom into the rails. The top is also fastened in this way, except that the screws are run through the rails into the top.

The two vertical pieces are now put in place. Drive nails through the bottom and into these pieces. On the top end use screws driven at an angle. Glue may be used if desired.

The doors are made to match these openings. The corners are mitered and the backs rabbeted to receive the panels. These panels may be made in art glass if so desired.

The horizontal piece for the drawer to rest upon is now put in place and fastened by driving nails through the vertical pieces. The drawer is made to fit this opening, and it should be lined with velvet to keep the silverware in good condition. [107]

The standards and shelves are put on as shown in the drawing. The mirror is put in a frame, which is made to fit the back opening and has the corners mitered and the back rabbeted to receive the mirror.

Thoroughly scrape and sandpaper all parts that are visible. The sideboard is now ready to be finished as desired.

A HALL OR WINDOW SEAT

Seat Made of Quarter-Sawed Oak

A simple design for a hall or window seat is shown in the accompanying sketch and detail drawing. Anyone who has a few sharp tools, and is at all handy with them, can make this useful and attractive piece of furniture in a few spare hours. [108] Quarter-sawed oak is the best wood to use in its construction, as it looks best when finished and is easy to procure. If the stock is ordered from the mill ready cut to length, squared and sanded, much of the labor will be saved. The following is a list of the material needed:

- 4 corner posts, 1-1/2 by 1-1/2 by 28 in., S-4-S.
- 2 side rails, 3/4 by 2-1/2 by 36-1/2 in., S-4-S.
- 2 end rails, 3/4 by 4 by 14-1/2 in., S-4-S.
- 2 side braces, 1 by 1 by 36-1/2 in., S-4-S.
- 2 end braces, 1 by 1 by 14-1/2 in. S-4-S.
- 1 seat, 1 by 16 by 35-3/4 in., S-4-S.

- 2 top end braces, 3/4 by 2 by 14-1/2 in., S-4-S.
- 6 slats, 3/4 by 2 by 6-1/2 in., S-4-S.

Square up the four posts and lay out the mortises according to the drawing. To do this, lay them on a flat surface with the ends square and mark them with a try-square. The tenons on the end and side rails are laid out in the same manner as the posts. The end rails should be marked and mortises cut for the upright slats as shown in the detail drawing. Fit the end and side braces with mortise and tenon joints.

The two end frames can now be glued and clamped together and set away to dry. Put all the parts together before gluing to see that they fit square and tight.

The seat should be made of one piece if possible, otherwise two or more boards will have to be glued together. The corners should be cut out to fit around the posts. It rests on the side rails and cleats fastened to the inner side of the end rails.

When the window seat is complete go over it carefully and scrape all the surplus glue from about the joints, as the finish will not take where there is any glue. Remove all rough spots with fine sandpaper, [109] then apply the stain best liked, which may be any one of the many mission stains supplied by the trade for this purpose. If this window seat is well made and finished, it will be an ornament to any home.

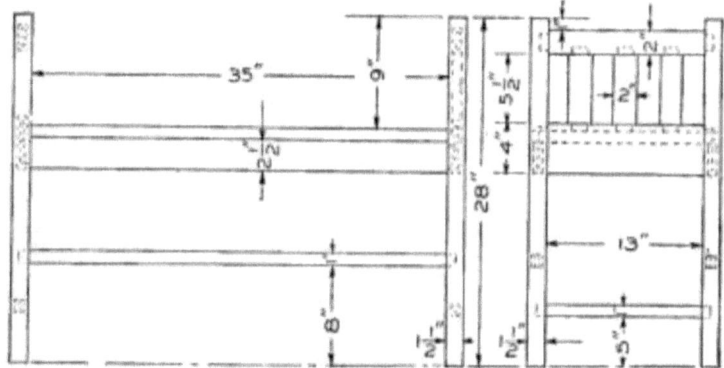

Detail of the Hall or Window Seat

A MISSION PLANT STAND

For the mission plant stand shown in the illustration secure the following list of quarter-sawed white-oak stock, cut and finished to size:

- 1 top, 3/4 by 15-1/2 by 15-1/2 in., S-2-S.
- 4 posts, 1-1/4 by 1-1/4 by 20 in., S-4-S.
- 4 rails, 3/4 by 3 by 11 in., S-2-S.
- 2 rails, 3/4 by 2 by 11 in., S-2-S.
- 1 shelf, 3/4 by 6 by 10 in., S-2-S.
- 4 slats, 1/4 by 2 by 12-1/4 in., S-2-S.
- 2 slats, 1/4 by 2 by 12-3/4 in., S-2-S.

Test all surfaces of the posts with a try-square to see that they are square with each other. Lay out the tenons on the ends of the rails as shown in the sketch and cut with a tenon saw and chisel. Arrange [110] the posts and rails as they are to stand and number each tenon and mortise. Lay out the mortises in the legs, taking the measurements directly from the tenon which is to fit that mortise. Cut the mortises, first having bored to the depth with a 1/4-in. bit.

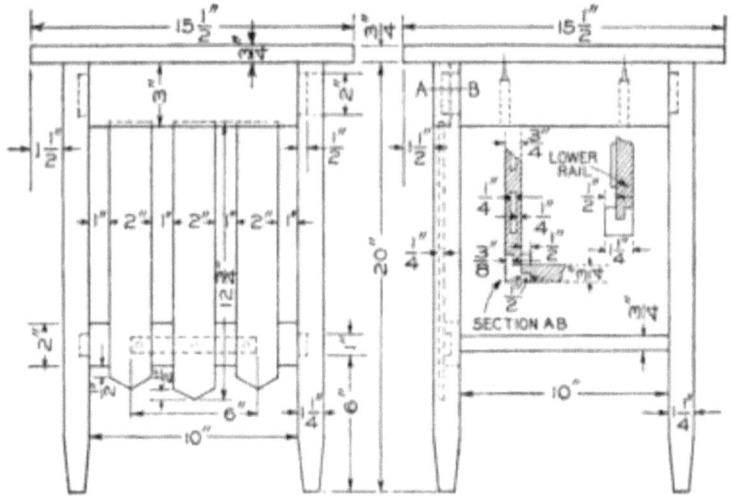

Detail of the Plant Stand

The slats should now be made and mortised into the top rail 1/4 in. They come outside of the lower rail and are held to it with two small brads, fancy-headed tacks, or round-head screws.

In laying out the mortises for the lower rails, care must be taken to have them set 1/8 in. farther in than the upper rails so the slats may come outside.

Set up the stand without glue or screws to see that all pieces fit accurately. Then glue up the sides [111] with the slats first. After these have set for 24 hours, fit in the other two rails and the shelf. Three flat-head screws should be used to hold the shelf in place. These must be placed so the slats will cover them when they are attached.

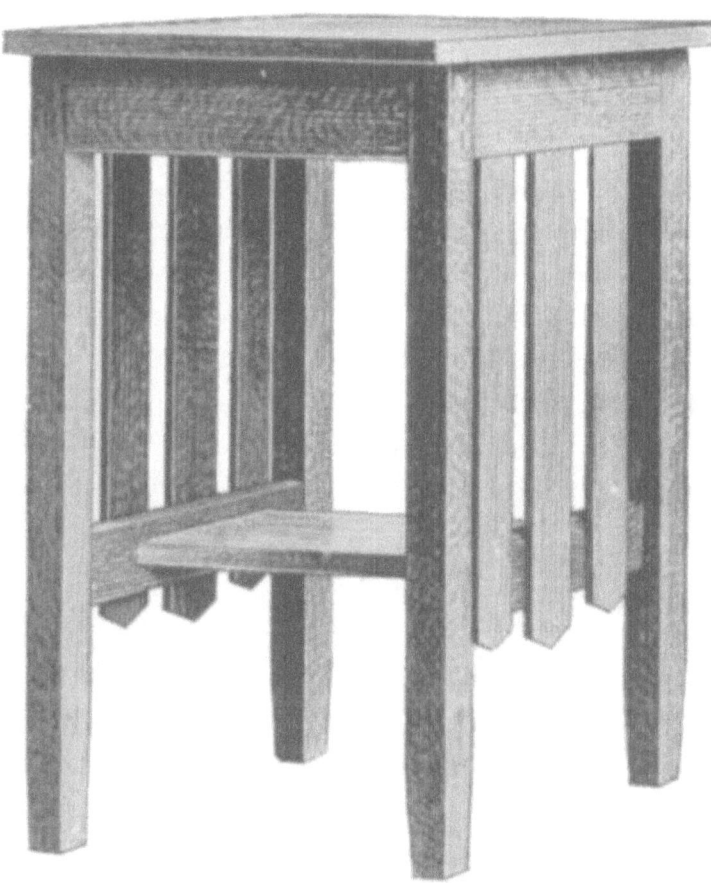

Complete Plant Stand

When this work is completed it is ready for the top. A good method of attaching the top is shown in the sketch. The screws used for fastening should be 2-in. No. 10. Bore into the rail 1-1/2 in. with a bit 1/16 in. larger than the head of the screw. Then bore through the rest of the way with a bit a little [112] larger than the shank of the screw. Thus a little space is left for expansion and shrinkage of the top.

Scrape and sandpaper thoroughly to remove all marks or glue spots. Finish with two coats of weathered-oak stain, followed by two coats of black wax.

A BEDSIDE MEDICINE STAND

The accompanying sketch and detail drawing show a design of a bedside stand. This is a very desirable piece of furniture and is simple and easy to make. Quarter-sawed oak is the best wood to use in its construction. The material should be ordered from the mill ready cut to length, squared and sanded. The following list of material will be required:

- 4 posts, 1-3/4 by 1-3/4 by 33 in., S-4-S.
- 1 top board, 1 by 19 by 19 in., S-4-S.
- 3 intermediate boards, 3/4 by 15-1/2 by 17 in., S-4-S.
- 2 side boards, 3/4 by 5 by 15-1/2 in., S-4-S.
- 1 back board, 3/4 by 4-1/4 by 14-1/2 in., S-4-S.
- 4 side rails, 3/4 by 2 by 16 in., S-4-S.
- 1 door, 3/4 by 9 by 14-1/2 in., S-4-S.
- 1 back board, 3/4 by 10-1/4 by 14-1/2 in., S-4-S.
- 2 panels, 3/8 by 9-1/2 by 15 in., S-4-S.
- 6 slats, 1/4 by 1 by 8-3/4 in., S-4-S.
- 1 drawer front, 3/4 by 4-1/4 by 14-1/2 in., S-4-S.
- 2 sides for drawer, 1/2 by 4-1/4 by 16 in., S-4-S.
- 1 back for drawer, 1/2 by 4-1/4 by 13-1/2 in., soft wood.
- 1 bottom for drawer, 1/2 by 13-1/2 by 15 in., soft wood.

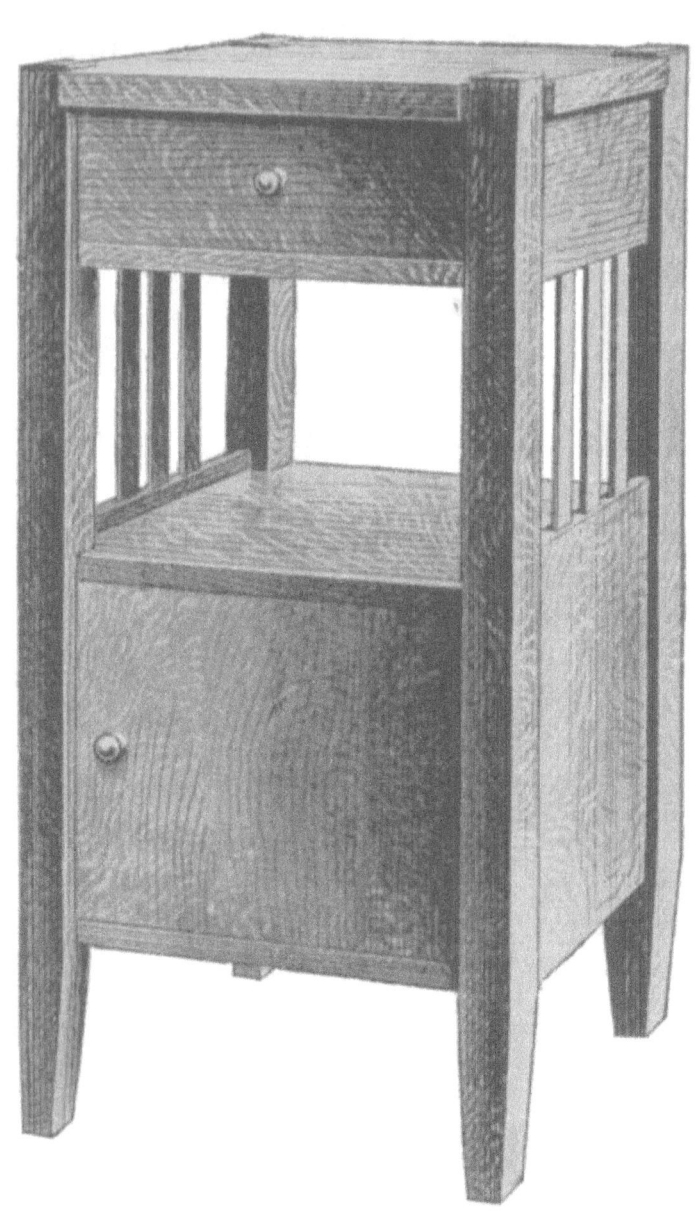

Medicine Stand in Quarter-Sawed Oak

Start work on the four posts by rounding the top corners and shaping the feet as shown. The four posts are identical and the mortises should be laid out on all four at once so as to get them all alike. These should be carefully cut with a sharp chisel. On the inner surface of each leg cut a groove to [113] hold the side boards of the lower compartment. Next prepare the two wide and the four narrow crosspieces, tenoning them to fit the mortises already cut in the legs. The lower crosspieces should also have grooves cut in them to hold the side boards of the compartment. The two complete sides can now be glued and clamped together and set away to dry. While they are drying the remaining parts of the stand can be made. The three horizontal boards are now made by notching out the corners to fit around the legs. They are supported by fastening small cleats to the inner surface of each crosspiece.

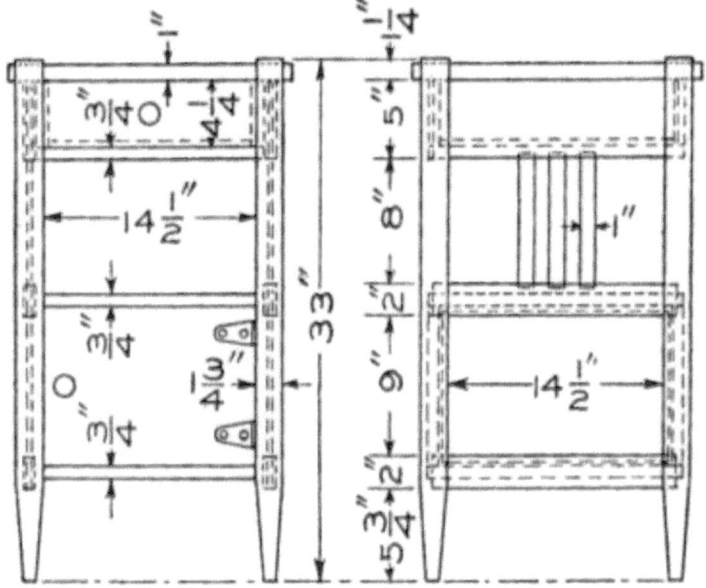

Detail of the Medicine Stand

The two ends can now be set up and connected. Notch out the corners of the top board and fit it in place. The top is fastened down by means of screws set in at an angle from below. The back boards can be of soft wood and are fastened in place in the [115] usual manner. The door should be of one piece if possible and should have suitable hinges and a catch.

Make and fit the drawer in place, and the stand is ready for the finish. First scrape all the surplus glue from about the points so the stain will not be kept from the wood. Finish smooth with fine sandpaper, then apply stain of the color desired.

A MISSION HALL CHAIR

This hall chair is designed to take up as little room as possible. For its construction the following stock will be needed:

- 1 back, 7/8 by 14 by 44 in., S-2-S.
- 2 sides, 7/8 by 14 by 17 in., S-2-S.
- 1 seat, 7/8 by 14 by 14 in., S-2-S.
- 1 stretcher, 7/8 by 6 by 16 in., S-2-S.
- 1 brace, 7/8 by 5 by 11 in., S-2-S.
- 1 piece, 7/8 by 7/8 by 44 in., for cleats.

These dimensions are for finished pieces, therefore 1/4 in. should be allowed for planing if the stock cannot be secured finished.

Lay out and cut the design on the back, sides, and brace. To cut the openings, first bore a hole near one corner to get the blade of a coping saw through and proceed to saw to the lines. Smooth the edges after sawing by taking a thin shaving with a sharp chisel. A file will not leave a good surface.

Mark the tenons on the ends of the stretcher and cut them with a backsaw and make smooth with a chisel. From the tenons mark the mortises in the sides through which they are to pass. [116]

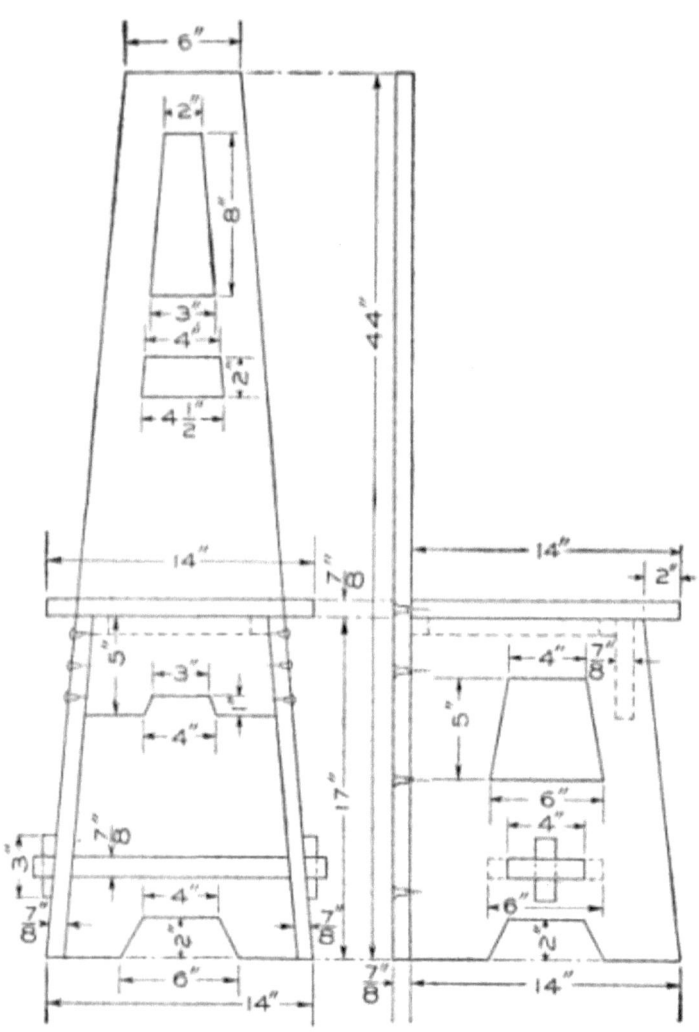

Detail of the Hall Chair

[117]

Complete Hall

Chair in Plain Oak

[118]

To cut these mortises, first bore a row of holes with a 5/8 in. bit, boring halfway from each side so as not to split off any pieces. Now make of scrap material the two keys and from them mark the small mortises in the tenons.

Before putting the chair together, the cleats for holding the seat should be fastened to the sides, back and brace. Use flat-head screws for this purpose. Then put the sides and stretcher together, and fasten the back to the sides with flat-head screws.

The brace should be put in next, using three round-head screws in each end. There only remains the top, which is held by screws through the cleats from the under side.

Stain with two coats of weathered or mission-oak stain, and then apply a thin coat of "under-lac" or shellac and two coats of wax.

www.ingramcontent.com/pod-product-compliance
Lightning Source LLC
Chambersburg PA
CBHW031414210526
45464CB00005B/1886